THE GREAT
EXTINCTIONS

**What Causes Them and
How They Shape Life**

Norman MacLeod

FIREFLY BOOKS

A Firefly Book

Published by Firefly Books Ltd. 2013

Copyright © 2013 Natural History Museum, London

First printing

**Publisher Cataloging-in-Publication Data
(U.S.)**

A CIP record for this title is available from the
Library of Congress

**Library and Archives Canada Cataloguing in
Publication**

A CIP record for this title is available from Library
and Archives Canada

Published in the United States by
Firefly Books (U.S.) Inc.
P.O. Box 1338, Ellicott Station
Buffalo, New York 14205

Published in Canada by
Firefly Books Ltd.
66 Leek Crescent
Richmond Hill, Ontario L4B 1H1

Cover design: Erin R. Holmes/Soplari Design
Interior designed by Mercer Design, London
Reproduction by Saxon Digital Services

Printed in China by C&C Offset Printing Co., Ltd.

Cover images
Front: ©Shutterstock.com/ixpert; holbox
Back: ©Shutterstock.com /Morphart
Creations inc.

This book was developed by:
the Natural History Museum,
Cromwell Road, London SW7 5BD

Contents

Introduction

FOR THE PAST 30 YEARS A SIGNIFICANT proportion of the scientific community has been obsessed with the idea of extinctions, especially the extinction of the 'dinosaurs' at or close to the boundary between the Cretaceous and Palaeogene intervals of Earth history. This interest pre-dates the current concern with the 'sixth' extinction, a hypothetical event that may occur in the future and which takes its name from the 'Big Five' ancient (mass) extinction events of the fossil record. The reasons for this and the sustained level of interest in extinction-related topics are many and varied. But they share a common source. The concept of extinction elicits a deep emotional reaction in most people today, in no small way because we all share an intuitive concern about transformations being wrought in our increasingly unnatural environment. When we see declines taking place in landscapes, animals and plants at the local, regional and even global scales we cannot help experience the sense of foreboding that comes from drawing obvious parallels between the status of our own species and the fates of other, far more ancient, species that 'ruled the Earth' in the distant past.

Much has been written about extinction. Many treatments of this topic end up claiming that the problem of understanding extinctions in general or particular extinction events has been solved (e.g. Raup 1991, Ward 1995, Alvarez 1997). In reality, the scientific community is far from having a detailed understanding of the enigma that is extinction, as attested to by the simple fact that 'extinction debates' constitute one of the longest-running scientific controversies in living memory. If a consensus regarding what 'killed' the dinosaurs, the ammonites, and their kin has been achieved (see Alvarez *et al.*, 1980, Schulte *et al.*, 2010), why do so many professional palaeontologists – especially those who know the extinction record best – stand outside it (e.g. see Archibald *et al.*, 2010)? Given the current state of knowledge about extinction as a phenomenon, what inferences for the contemporary and future management of our planet can, or should, we draw? What type of cataclysm does it take to extinguish 50, or 60, or 75, or 90% of all species on the land and in the sea, as has happened repeatedly in the Earth's distant past? What causes the sort of changes in the environment that drive extinction rates to these astonishing levels and over what timescales? Perhaps most importantly, how does a planet recover from devastations of such magnitude?

I have undertaken and published extinction research using the fossil record as my primary source of data for most of my professional career. I and my colleagues have grown up (literally) with this research programme, this scientific debate, this public controversy. Like all participants in any human activity, I have a particular point of view that I believe conforms to the most reasonable interpretation of the greatest proportion of evidence currently to hand. I disagree with explanations offered by some of my colleagues and some of them disagree with me. Such is the character of healthy scientific debate. But my goal in this book is not to simply present the case for my own point-of-view by citing the evidence in its favour and ignoring contrary observations. Rather, it is to present the data extinction researchers of all persuasions work with as fairly as I can, mentioning all the nuances, caveats and assumptions that often get left out of presentations for a popular audience. Once this evidence has been presented it will be up to you, the reader, to come to your own conclusions about extinctions, what has happened in the past, and what might occur in the future. No doubt my own biases will creep in from time to time. This is inevitable. I pledge here to make a diligent effort to identify instances in which I am offering a personal opinion or interpretation. More than this though, I hope to convey some inkling of the excitement, the novelty, the frustration and the sense of grandeur that accompanies the study of one of nature's most common processes, but also one of its deepest mysteries.

It has been said that the secret to a long life is to have a chronic incurable disease and to keep treating it. By the same token, the secret to a productive life in science is to have a chronic insoluble problem and to keep working on it. By this measure I and my extinction-studies colleagues on all sides of the interpretational fence have been very fortunate indeed.

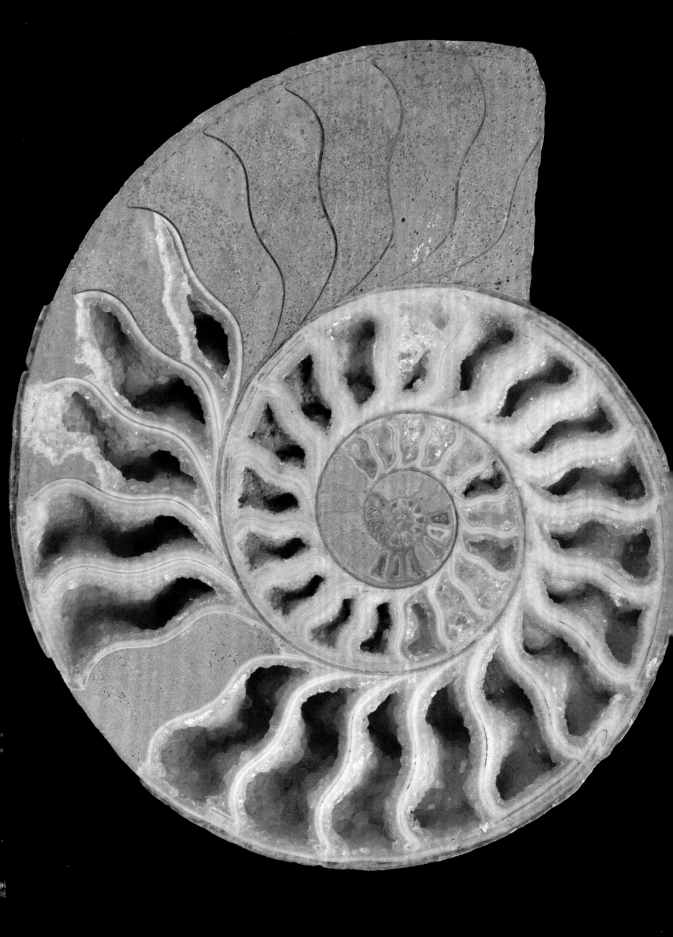

1 What is extinction?

EXTINCTION OCCURS WHENEVER the last individual that can be assigned to a taxonomic group (e.g. species, genus, family) dies. This usually comes about as a result of an ecologically long period of declining numbers of individuals and a progressively contracting geographic range. Ecologists and demographers often use the term 'extinction' to refer to the disappearance of a group from a local area or region. The correct technical term for a local or regional disappearance is extirpation. Palaeontologists and conservationists usually use the term extinction in its correct technical sense, to refer to the disappearance of the group globally. Owing to the vagaries of the fossil record, however, it is often very difficult to know whether the disappearance of a fossil taxon represents an instance of extirpation or true extinction. Irrespective of these difficulties, extinction plays an important role in driving many natural processes forward. Indeed, along with the core biological processes of adaptation, selection and speciation, extinction is fundamental to understanding the history, the present state, and the future of the natural world.

From a mathematical point of view, extinction comes about when the average birth rate of an established population remains less than its average death rate for a sufficiently long time interval to allow random fluctuations in the yearly birth and death rates to reduce the population size to zero. These fluctuations are caused by a wide range of independent factors such as environmental change, introduction of predators, introduction of competitors, elimination of critical resources (e.g. food, sheltering sites, nesting sites) introduction of diseases, etc.

The inter-relations between factors that promote and/or discourage extinction are often studied using mathematical models to estimate the probable time to extinction for a series of populations of different initial sizes. For example, assuming the factors that lead to extinction are random and constant over time, the expected time to extinction can be modelled by the following equation:

$$p_0(t) = (dt/1 + bt)^i$$

where: $p_0(t)$ = probability of extinction
b = birth rate
d = death rate
t = time to extinction (in reproductive cycles)
i = initial population size

OPPOSITE The Middle Jurassic ammonite *Brasilia bradfordensis*. This specimen has been sectioned to show its internal chambers separated by partitions or septa. During fossilization some of these chambers were filled with mud completely, some only partially, and some not at all. Disappearance of the ammonites is traditionally associated with the end-Cretaceous mass extinction, but ammonite fossils have recently been reported from sediments that overlie (i.e. are younger than) the end-Cretaceous extinction horizon in Denmark.

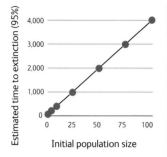

Estimated time to extinction (95%)

Initial population size

ABOVE Results of a series of estimated time to extinction (in generations) simulations for populations of different sizes under the condition that birth and death rates are equal and have the value of 0.50. Experiments like this show how extinction susceptibility is linked to population size.

If we set the probability of extinction at 95% and the birth and death rates to the same constant values (say 0.50) we can examine the behaviour of the system represented by the equation for a hypothetical set of populations of various sizes by graphing the results (see left). Note that a relatively small increase in the initial population size has a large effect on the estimated time to extinction. This simple mathematical experiment shows that, when population sizes are small (e.g. less than 100), extinction can happen over very short time intervals, even when the range of environmental change remains constant.

The graph shown left predicts that species with large populations should be more resistant to extinction than species with small populations. Is this an accurate description of what happens in nature? We could set up an experiment in a laboratory to test our model. But laboratories are not natural, by definition. Fortunately, this experiment has already been performed for us under totally natural conditions – on islands.

The study of island species has proven to be one of our best sources of information about extinction as a process. Island species not only exist at smaller population sizes than continental species, but the sizes of island populations are set by the sizes of the islands themselves. Data on island bird extinctions confirm the expected relationship. Of bird species that are known to have become extinct over the last 300 years, the overwhelming majority were indigenous to islands the size of New Zealand or smaller.

This relationship also holds for comparisons among islands. On the Channel Islands off the California coast between 1917 and 1968 as many as 70% of the bird populations on each island became extinct. The highest levels of extinction were recorded from the smallest island (70%) in the group and the lowest levels (36%) from the largest.

Comparable figures are now available from many islands scattered around the world. While there is some concern that these data represent overestimates owing to inadequate historical records and/or sampling, the association between population size and resilience in the face of extinction is clear. Most ecologists agree that population size is the most significant single factor in determining extinction susceptibility.

Given that each species' inevitable fate is to become extinct it is fair to ask whether there are similarities and differences between species that bear on the question of extinction resilience. Do the environmental processes that strike species down operate randomly with respect to each species' evolutionary history, or is there an underlying pattern that might help us understand which species might be at more risk of extinction than others? Interestingly, the answers to this question appear to depend on the timescale considered relevant.

In the modern world the antithesis of a species that is becoming extinct is one whose population size (and usually its geographic range) is increasing. Since virtually all species are thought to begin as relatively small, localized populations, the interval between speciation and extinction can be likened to a developmental cycle in which a series of phases can be recognized. This taxon cycle has been most extensively documented on islands where historical records can be used to trace the dates of species' arrival at, and disappearance from, specific geographic localities. By constructing historical maps of species' geographic distributions over time the classical taxon cycle stages of expanding (stage I), differentiating (stage II), fragmenting (stage III) and endemic (stage IV) distributions can be recognized. While taxon cycle maps are particularly easy to construct for island biotas, the concept is applicable to continental and even marine settings.

Given the taxon cycle, it might appear as though species that have a long evolutionary history would be, on average, more resistant to extinction than a species whose history is short. In the early 1970s the evolutionary biologist Leigh Van Valen studied this question by consulting the technical literature on fossils and looking up the time intervals over which 24,000 fossil taxa existed. Van Valen summarized these data as a series of 'survivorship' graphs that show the distribution species durations. Examples of survivorship graphs from the study are shown below.

When these graphs were compared a number of interesting features became apparent. First, different types of plants and animals appear to have characteristically different extinction susceptibilities. For example, Van Valen's data indicated that the average duration for bivalve species is 5 million years while that of a mammal species is just 1 million years. It is by no means clear why these differences between groups of organisms exist. But exist they do.

Even more surprisingly, within various higher taxonomic groups (families, orders) the proportion of genera and species that become extinct per time interval appears to be nearly constant. Van Valen interpreted this constancy to mean that evolution does not operate in such a way as to grant long-lived species greater extinction resistance. If these data are taken at face value, over time species do not, on average, get any better at avoiding extinction. Rather, over the whole of geological history, a set proportion of species have become extinct in any given time interval, though this number differs from group to group. Subsequent researchers have disputed aspects of Van Valen's interpretation. However, his basic findings have stood up to repeated criticisms remarkably well over time in that they have predicted the results of new research inspired by Van Valen's critics. This process of attempting to falsify the interpretations of colleagues by comparing predictions with data drawn from nature is how science progresses. But to understand the potential causes and implications of Van Valen's findings we must first understand the role extinction plays in evolutionary processes.

BELOW Taxonomic survivorship curves for fossil genera of foraminifera (left), bivalves (centre) and mammals (right). Analysis of the stratigraphic ranges of many fossil groups indicate that evolutionary durations are highly specific to each taxon but that over long intervals of geological time extinction susceptibility is nearly constant. (See Van Valen 1973.)

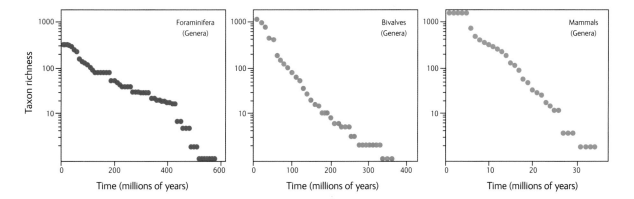

2 Evolution, fossils and extinction

THE CONCEPT OF EXTINCTION has had an odd relationship with that of evolution. Darwin's view of extinction was that it represented the natural outcome of the struggle for existence that lies at the heart of his theory of natural selection. Under this theory individuals compete with each other for resources located in natural environments (e.g. food, shelter, nesting sites, mates). Individuals that share some morphological, physiological, behavioural or other characteristic that enables them to win this competition more often than other individuals will have a greater likelihood of surviving to reproductive age and/or will produce more offspring than those who do not. If the traits responsible for this performance difference can be passed from parents to offspring through genetic inheritance, individuals possessing traits linked positively to survival will become more common in local populations undergoing selection.

Provided the traits conferring this competitive advantage spread throughout a local population the potential exists for the members of this population to lose the ability, or the opportunity, to interbreed successfully with members of the other populations belonging to the species. At the point where this reproductive disruption occurs a new species is formed. Subsequently, both ancestral (parent) and descendant (daughter) species will proceed along different evolutionary trajectories.

OPPOSITE **Charles Darwin (1809–1882)** formulated the theory of evolution via natural selection. Darwin acknowledged the reality of extinction, but did not accept the idea of mass extinction and did not feel extinction played a creative role in evolution.

EVOLUTION AND EXTINCTION

The mode of evolutionary change in which both parent and daughter species persist is termed cladogenetic evolution (see p.12). Once cladogenetic evolution is complete the daughter species enters into competition with all other species in the local area – including the parent species. As a result of the struggle for limited environmental resources the possibility exists that, over time, the overall size of the daughter species or one or more of its competitor species would be reduced to the point at which either local extirpation or global extinction is likely due to random population size fluctuations.

Alternatively, if the traits conferring this competitive advantage spread throughout all populations of a species it is possible that, over time, the entire species will change its fundamental character. When this occurs taxonomists – scientists who study and name species – have recognized the change by giving the changed species a new name, to distinguish it from the older form of the species that existed at an

RIGHT Diagrammatic examples of (a) anagenetic and (b) cladogenetic evolution. Under anagenetic evolution an entire population or species undergoes a change in its form and genetic make-up in response to natural selection. Taxonomists recognize this change by giving the new form a different name. Under cladogenetic evolution only part of the species, usually an isolated population, undergoes this evolutionary transition with the ancestral species persisting in other areas.

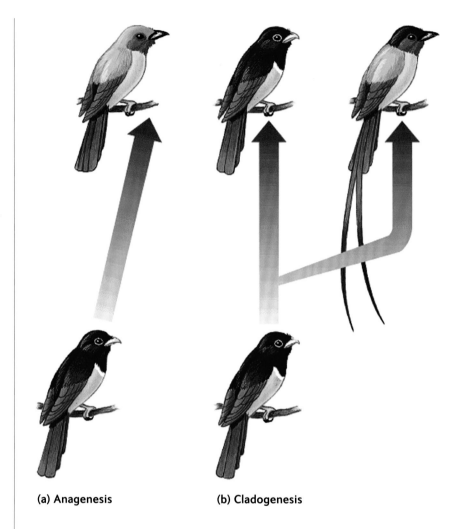

(a) Anagenesis   **(b) Cladogenesis**

earlier time. The naming of new species in this manner is especially common among palaeontologists who need to distinguish between older and younger forms of an evolving lineage, both forms of which are preserved in the fossil record. This mode of wholesale population change is termed anagenetic evolution (see above).

Anagenetic evolution differs fundamentally from cladogenetic evolution insofar as, under anagenetic evolutionary change the parent species is obliged to disappear from the biological record. But unlike true extirpation and extinction, this disappearance does not result from the fact that the species failed to cope with environmental change. In fact, the disappearance of the parent species in an anagenetically evolving lineage is a signal that the lineage is coping with environmental change successfully, by changing its character through adaptation. Because the implications of apparent species loss due to anagenesis are so different from the implications of real species loss due to competition, the former is termed a pseudoextinction event, or pseudoextinction. In reality, all that goes extinct in a pseudoextinction event is the name of the ancestral species.

In the cases of both cladogenetic and anagenetic evolution, and for a variety of reasons, Darwin viewed extinction as a gradual process that comes about as a result of natural selection according to the mechanisms described above. In particular, he opposed the idea that large numbers of extinctions were caused directly by physical processes in the form of unusual or catastrophic natural events. This was in keeping with the theory of gradualism in natural processes to which many of Darwin's philosophical and scientific colleagues also subscribed (e.g. the geologist Charles Lyell) and that many of those who opposed Darwin's theory of natural selection rejected (e.g. Sir Richard Owen). As a result, the topic of extinction does not figure greatly in Darwin's seminal book on evolution, *On the Origin of Species* (1859), or indeed in any of Darwin's books.

Darwin's views on species extinction went on to become scientific orthodoxy throughout the late 1800s and early 1900s. By the 1940s and 1950s, the scientific community's view of extinction had drifted from that of regarding evolution as a by-product of the struggle for existence to being quite a passive process in which the ecological roles vacated by extinct species might lay fallow for (literally) tens of millions of years before natural selection was able to fashion a new occupant for the former role. For example, in his book *Tempo and Mode in Evolution* (1944), the influential vertebrate palaeontologist George Gaylord Simpson pointed to the fact that, according to the fossil record, millions of years separated the vacation of the marine, air-breathing predator role once occupied by ichthyosaurs and its reoccupation by toothed whales. Similarly, tens of millions of years passed between the demise of the non-avian dinosaurs as the dominant large terrestrial vertebrates and the appearance of a similarly diverse assemblage of large mammals. On the whole Simpson and others of his time regarded extinction as playing even less important a role in evolution than that envisioned originally by Darwin. One could lament the passing of such magnificent creatures as dinosaurs, pterosaurs, ichthyosaurs, plesiosaurs, sabre-toothed cats, mammoths and mastodons from the scene. But informed scientific opinion throughout the 1930s, 1940s and 1950s was that extinctions were incidental by-products of normal natural selection and that the biosphere seemed to take its time replacing groups that became extinct.

The twenty-first century view of extinction is decidedly different. Beginning with the publications of Otto Schindewolf in Germany in the 1950s, and Norman Newell in US in the 1960s, catastrophism – the idea that the Earth has been repeatedly subject to massive environmental convulsions – was rehabilitated as a feature of the natural environment. Previous generations of earth scientists, including Darwin and Simpson, had rejected nineteenth-century catastrophist theory because of its appeal to vague and mysterious processes that were, by definition, unavailable for scientific study. [Note: catastrophism was also rejected for political (association with the French Revolution) and religious (association with biblical fundamentalism) reasons.] However, by the 1960s palaeontological evidence for the coordinated loss of major animal groups over relatively short intervals of geological time was such that Lyell's and Darwin's ideas about such patterns being accounted for by generalized appeals to imperfections in the fossil record were no longer credible.

Similarly, by the 1970s the physical evidence for massive volcanic eruptions (much larger than had ever occurred in human history), profound changes in sea level, and even whole continents drifting across the surface of the Earth, all having occurred within surprisingly short geological time intervals, became too compelling to ignore in terms of their implications for the history of life.

The late twentieth century reappraisal of the significance of relatively short-term, often violent, processes in Earth history has often been portrayed as a neocatastrophist revolution in the earth sciences. In hindsight, this advance in our understanding of the natural world is perhaps more accurately regarded as an extension of the doctrine of actualism, the idea that processes operating in the present time have also operated in the past. In Darwin's day the idea of an asteroid colliding with the Earth or a volcano erupting tens of thousands of cubic kilometres of lava was all but inconceivable. But by the 1960s evidence had accumulated that, throughout geological history, the Earth had in fact been subjected to far more violent and far more long-lasting environmental disruptions than those that had been recorded in human history.

BELOW The Barringer Meteor Crater, near Flagstaff, Arizona. This structure was once thought to be created by a volcanogenic steam explosion. Confirmation of its true origin, the result of a meteorite impact, was made by Eugene Shoemaker in 1960 with the discovery of its association with rare forms of quartz that are only produced by severe pressure shocking. This crater was made by a nickel-iron meteorite c. 50 m (154 ft) wide.

An outstanding example of the culmination of this trend to extend the range of processes considered to be 'natural' was the identification of large circular depressions of the Earth's surface (see p.14) as being craters left by meteor collisions similar to those that mark the surface of Earth's moon. Previously these features (on the moon and the Earth) had been thought to be the products of volcanic processes. The case for an extraterrestrial origin for these geological features was finally made by Eugene Shoemaker, who found rare high-pressure forms of the mineral quartz in and around what later came to be know as the Barringer Meteor Crater in the US state of Arizona. Experimental evidence indicates these unusual forms of quartz can only be formed under conditions of intense physical shocking such as would be expected to result from the impact of a sizeable meteor on the surface of the Earth. Once such rare but large-scale processes as meteor impacts and massive volcanic eruptions (see below) had been admitted to the pantheon of possible causes of global change, it was but a small step for palaeontologists to begin to wonder about the implications these processes held for the history of life.

The (re)admission of catastrophic physical processes to the list of possible extinction mechanisms also meant evolutionary processes that did not involve direct competition between species must now be considered possible, at least in principle. Since species have definite, and surprisingly short, lifespans, they cannot develop adaptations to cope with rare and violent environmental events that have never occurred during their own, or their direct ancestors', existences, the likelihood of species surviving such events, then, could come down to a matter of random chance: occupation of an environment or region that just happens to escape devastation; an environmental tolerance that just happens to promote persistence under unusual conditions; an ability to switch to a set of ancillary resources that just happen to be available in the wake of a catastrophe.

This scenario predicts that a class of extinction-inducing mechanisms might exist that could, in principle, nullify any competitive advantage a species might possess due to Darwinian adaptation. Rare and violent environmental events might have profound effects on the course of evolution, effects that would come about as a result of forces originating from outside the range of normal evolutionary processes. To paraphrase the noted palaeontologist David Raup, the significant aspects of the history of life could be the result of good luck instead of good genes.

This 'bad luck' model of extinction susceptibility can be used to understand one reason why the survivorship curves constructed by Van Valen indicating species extinction rates appear to be constant. Over short periods of geological time natural selection may produce extinction-resistant species, but these species might be eliminated from evolutionary systems by either single or unusual combinations of rare events, that induce massive disruptions to ecosystems such that species are eliminated randomly, without reference to previous successes in coping with 'normal' instances of environmental change and/or inter-species competition - the types of changes encountered by species as a result of normal, Darwinian natural selection. Such rare events would lie at the extreme end of a spectrum of environmental challenges faced by species over the short, medium and long terms. But in system terms, the dynamic character of environmental challenges that arise at all levels, coupled with the unrelenting (Darwinian) need for every species to compete with every other species (to a greater or lesser extent) in its immediate environment for resources critical to its survival, means that no species can ever develop a competitive advantage that can be sustained over the long term. In other words every species must 'run' (in an adaptive sense) as fast as it can simply to remain in the same competitive place with regard to all other species in its local environment. Because of its allusion to an observation make by the Red Queen in Lewis Carroll's story , this explanation of extinction constancy has been termed the 'Red Queen' model of evolutionary change, which Van Valen likened to a 'new evolutionary law'.

Most of this book will be devoted to detailing – albeit briefly – the results of this reappraisal of the range of physical processes that might result in the random extinction of ancient organisms along with the implications this reappraisal has for our understanding of the effects of changes in the global environment induced by human activities. In particular I will be delving into whether a single class of very intense and unusual events are primarily responsible for the great geological extinction events or whether these are more reasonably associated with unusual juxtapositions of events that perturb the global environment at lower levels of intensity, but over longer intervals of time. To distinguish between these two scenarios I'll refer to the former as the single cause (SC) scenario and the latter as the multiple interacting cause (MIC) scenario.

THE FOSSIL RECORD

Since primary data for studying the causes and effects of very large extinction events come from the fossil record, we need to understand the nature of that record and its constituents. Fossils are the petrified remains of ancient organisms. Fossilization is a natural process that begins when a dead organism's remains – including the impressions or marks it made on its surroundings and/or the chemical traces it left in its local environment when alive – are buried by sediments. Once isolated from the surface environment the body and/or traces of the organism can, when conditions are favourable, be either preserved in their original form or (more commonly) replaced by minerals that preserve the form of the original body part or mark left by an organism's activity. As a result, fossils are resistant to further physical change. In this way a record of the original organism's morphology, either in whole or in part, may be preserved for long periods of time; in some cases literally for billions of years.

Because the physical and chemical processes that lead to fossilization have been operating for as long as life has been present on the Earth, many of the fossils that have been entombed in the Earth's sedimentary rock record represent extinct species. In some cases the quality of a fossil's preservation is such that it may be compared directly with living species, whose hard-part morphology is familiar to modern scientists, and which can be investigated in as much detail as modern technology allows. If the fossil is indistinguishable from the form of a living species it is generally regarded as representing an individual of the same species, but one that lived at an earlier time in Earth history. However, if the fossil has a unique form, or if there is some unique detail of the fossil's overall form that is not found among the set of species known to be alive today, the fossil is assigned a new species name and regarded as having belonged to a species that existed at some former time, but has since become extinct.

Once a fossil has been identified the time interval over which it existed can be estimated as the span between the earliest and latest recorded observation of specimens assignable to the species. This span may be given in terms of a physical distance in a stratigraphic section, a relative date determined via reference to the geological timescale, or an absolute date estimated usually using radioisotopically datable minerals found in association with or near to the specimen. In some cases the absolute dates of a fossil species' occurrence can also be estimated by radioisotopic calibration of the time-interval boundaries that comprise the geological timescale (see below).

How many fossil species are there? In a sense this is unknowable as hundreds to thousands of new fossil species are discovered by palaeontologists, and amateur fossil enthusiasts, each year. Nevertheless, estimates can be made.

Table 1. Estimates of species lifespans for various taxonomic groups based on data from the fossil record.

Group	Average longevity in millions of years)
All mammals	1
Caenozoic mammals	1–2
Diatoms	8
Dinoflagellates	13
Planktonic foraminifera	7
Caenozoic bivalves	10
Echinoderms	6

Based on the average lifespan duration for fossil species (Table 1) it can be concluded, with a good deal of confidence, that the proportion of currently living species is less than 1% of all species that have ever lived during the whole of Earth history. In other words, the fossil record represents the only source of direct evidence available for the overwhelming majority of species that have inhabited our planet over the course of the last 3.4 billion years.

This having been said, the fossil record is far from a perfect record of this pantheon of past life. Realistically, palaeontologists can harbour no hope of ever being able to obtain a complete record of soft-bodied organisms, especially if these organisms are small and/or inhabited ephemeral environments (e.g. a forest, a riverbed).

The fact that a substantial proportion of modern biodiversity falls squarely within this category is cause for concern when making many types of palaeontological interpretations. Nevertheless, biologists and ecologists know surprisingly little about these sorts of organisms even today. Advances in our knowledge of them will surely improve, largely through DNA barcoding, which will lead to a better understanding of their taxonomy and phylogenetic relations. This will, in turn, enable a host of other investigations to be undertaken. What we can do at present, however, is focus on that component of the modern biota whose bodies include materials likely to serve as the basis for fossils (e.g. shells, teeth, bones, scutes, spines). Of these there are approximately 250,000 living species.

By making various assumptions about standing biodiversity in the past and rates of species turnover based on the fossil species already described we can estimate there may be as many as 12 million species that could potentially be recovered from the fossil record. Of these, perhaps 500,000 fossil species have been found and described to date. This suggests that as little as 5% of all species that might be included in the fossil record are presently known to science. Of course, some groups have much more complete fossil records than others. For example, it is very unusual for a new species of Pleistocene planktonic foraminifera (a group of single-celled zooplankton) to be described in the palaeontological literature, whereas new fossil dinosaur species are discovered with astounding regularity. Indeed, a depressingly large proportion of all dinosaur species are represented only by single specimens, and often fragmentary specimens at that.

Just as important, the fidelity of the fossil record varies from organismal group to organismal group. For example, marine organisms encased in hard shells (e.g. ammonites, brachiopods) exhibit very good, high-fidelity fossil records, whereas marine groups whose bodies contain only small teeth, spines and scutes as hard constituents (e.g. worms, sharks and rays) do not (see below).

RIGHT Estimates of the quality of the fossil record for various marine invertebrate and vertebrate groups. (Redrawn from Foote and Raup 1996). Note the distinction between the fossil records of Cephalopoda (squid, octopus, ammonites, upper left) and Chondrichthyes (sharks and rays, lower right). Very few living cephalopod families are known from the fossil record because they lack a hard, robust external shell. However, since many ancient cephalopods did possess a hard, robust external shell (e.g. ammonites), the fossil record of ancient cephalopod species is thought to be reasonably complete. Conversely, few modern shark and ray species have been recovered from the fossil record and, owing to their relative lack of hard parts the fossil record of these species is expected to be of poor quality. The quality of the fossil record of most organismal groups falls between these two extremes.

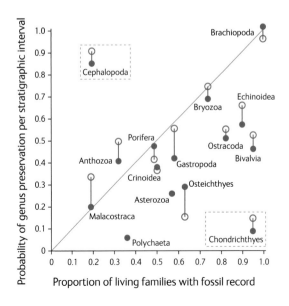

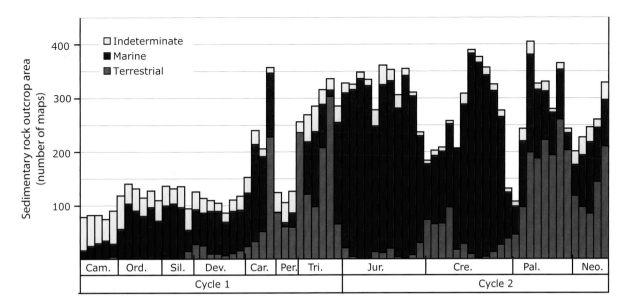

The amount of sedimentary rock outcrop area exposed at the Earth's surface also differs among intervals of varying ages (see p.19 top). Generally speaking, younger time intervals are represented by a greater proportion of outcrop area than older time intervals. This means that, all other factors being equal, the fossil record of younger time intervals is likely to be more complete than that of older time intervals.

These differences between taxonomic groups and age cohorts with respect to the quality of their fossil records need to be kept in mind throughout our discussion of the natural history of extinction. Often, the fossil record leaves much to be desired a fact emphasized by many biologists including Charles Darwin who was frustrated that it did not provide more direct support for his theory of evolution via natural selection. But the fossil record can also be a surprisingly complete, high-fidelity window through which we view the history of life on Earth. Above all the fossil record challenges us to be creative in our hunt for clues to the Earth's past and, not infrequently rewards this creativity with breathtaking detail. Regardless, and despite all its imperfections, the fossil record remains the best – indeed the only – direct record we have of the character, the form, the victims and the survivors of the great extinction events, as well as the only reliable guide to the manner in which the biosphere has coped with massive environmental changes throughout Earth history.

ABOVE Surface area of sediments of different ages and environments outcropping in Spain, France, England and Wales. Note the generally increasing trend in average outcrop area for younger time intervals in the marine record and the gap in outcrops of Middle Mesozoic age (a time of high sea-levels) in the terrestrial record. Note also the relative lack of fossil deposits of any type in the Late Permian (Per.) and Late Cretaceous (Cre.). Systematic differences in the probability of finding fossils of particular ages bias our estimates of species richness and extinction. (Redrawn from Smith and McGowan 2007.)

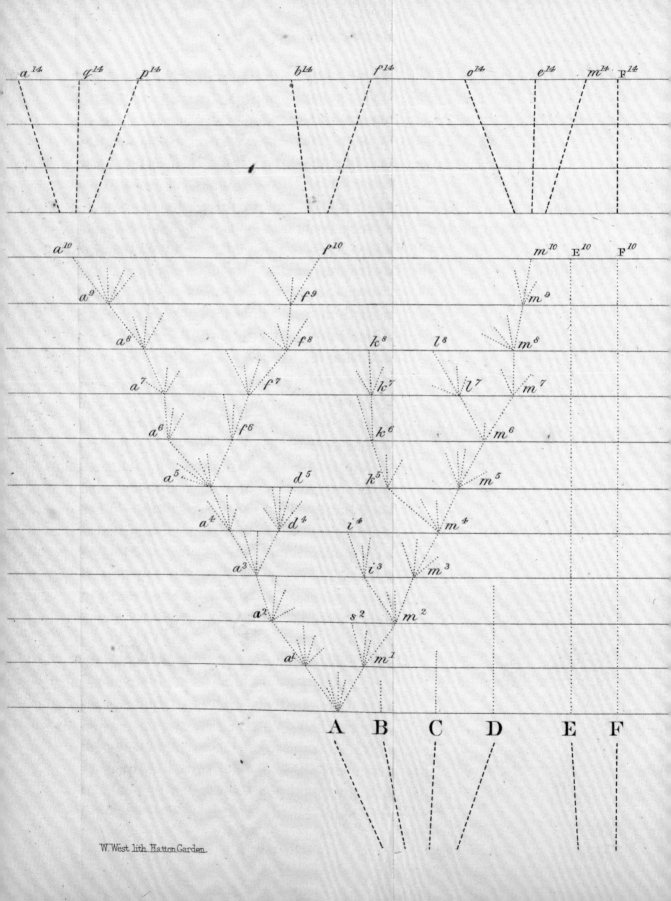

3 Patterns in extinction data

THE DATA RESEARCHERS USE TO STUDY the great extinction events come from the fossil record. But the data of fossil records constitutes more than just the fossil remains themselves. When palaeontologists collect fossils they are not only interested in the fossil object. They also record detailed information about where the fossil was collected from, both in terms of the geographic location of the rock outcrop or, in the case of drill cores, the drill hole, and the position of the fossil in the vertical sequence of rock layers. These data are not collected simply to allow the palaeontologists to find more fossil specimens. They are as important primary observations as the identity of the fossil itself. The geographic location of the fossil establishes its location on the Earth's surface. Owing to the processes of sea-floor spreading and continental drift (see below), and the long time interval that separates the fossil's original deposition and its discovery, the geographic location of its collection is often quite different from the location – and the environment – in which the plant or animal actually lived. The fossil's position in the local rock sequence is even more subtle and intriguing for this information establishes the fossil's position in time. Fossils that occur at different levels within a rock sequence are not just occurring in different places. They are occurring at different times in Earth history, often considerably different times. When palaeontologists look at a rock sequence they don't just see a pile of rock, but the pages of a history written by nature herself.

These three observations – the fossil object, its geographic position in space and its position in time – are the primary data of the fossil record. They are the data used to study the great extinction events. When authors from Darwin to the present day refer to the 'imperfections of the fossil record' they are not only referring to the fact that not all ancient species have been preserved as fossils, but also to the fact that the fossil record is both complex and incomplete in terms of its representation of geography and time. Whole regions of the Earth's surface are not part of the accessible fossil record because they are buried beneath other rocks, because no sediments were deposited at that place and time, or because the sediments that were deposited have been destroyed by a host of natural processes (e.g. erosion, metamorphism, subduction). Similarly, at any given geographic locality whole intervals of geological time may not be accessible, may not have been represented by sediments, or may have been destroyed. Perhaps the best way to think about the fossil record is that it resembles a book that's been torn apart and the various chapters scattered about the landscape. It is the palaeontologist's task to find as

OPPOSITE The tree of life engraving from Charles Darwin's, *On the Origin of Species*, 1859. This was the only diagram included in Darwin's great book on the subjects of evolution and natural selection and illustrates his view of evolution as a process that involves both cladogenesis and anagenesis.

many chapters as they can and to reconstruct the book, putting the chapters back into their original sequence, identifying the missing sections and trying to infer what was once there, but has been lost, on the basis of the information recovered to date. The complexity of this task needs to be kept in mind when thinking about what we know, and indeed what we can know, about the great extinction events.

SEA-FLOOR SPREADING AND CONTINENTAL DRIFT

As far back as 1596 geographers have been intrigued by the reciprocal similarity of the shapes of the eastern coast of South America and the western coast of Africa. This circumstantial correspondence led some to believe these two continents, and perhaps others, had been joined together at some time in the Earth's past. The meteorologist and polar explorer Alfred Wegener developed this theory most fully in 1912, but his efforts were preceded by a long trail of similar publications by others. In addition to the map similarities Wegener pointed to similarities in the shapes of the continental shelf margins, rock types, geological structures and fossils as evidence for his theory of *Kontinentalverschiebung*, or continental drift (see below).

Many modern histories of geology argue that Wegener's ideas about continental drift foundered on the issue of mechanism. Wegener argued that the continents floated in a sea of denser, basaltic ocean crust that, over long periods of time, behaved as a viscous fluid. He thought the continents plowed their way through this fluid as a consequence of tidal forces and/or gravity, deforming it into mountain ranges on the continental margins in the process. Others suggested the continents slid laterally

RIGHT Some of Alfred Wegener's evidence for his theory of *Kontinentalverschiebung*. See text for discussion.

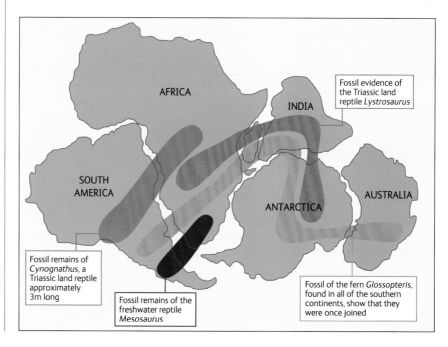

AFRICA

INDIA

Fossil evidence of the Triassic land reptile *Lystrosaurus*

SOUTH AMERICA

ANTARCTICA

AUSTRALIA

Fossil remains of *Cynognathus*, a Triassic land reptile approximately 3m long

Fossil remains of the freshwater reptile *Mesosaurus*

Fossil of the fern *Glossopteris*, found in all of the southern continents, show that they were once joined through the basaltic fluid following regional uplift. Notably in the later 1920s and early 1930s the British geologist Arthur Holmes suggested heat formed in the Earth's mantle as a by-product of radioactive decay would set up convection currents in the mantle that would rise from the Earth's core and spread laterally across the surface of the planet's mantle. Under this model the continents rode passively on top of the laterally migrating mantle. Interestingly, this is, in essence the mechanism accepted today for the successor to Wegener's continental drift theory, the theory of plate tectonics. However, sea-floor spreading was not accepted by most geologists until the 1960s. This means the traditional histories are wrong. A plausible — indeed so far as we know the correct — mechanism had been proposed by the mid-1930s. Yet at that time continental drift, like catastrophism, was relegated to the category of 'disproven' ideas by most practising Western geologists, geographers and physicists.

As Oreskes (1988, 1999) has shown, resolution came, as it often does in science, not through the postulation of a new mechanism, in this case to drive the continents laterally across the Earth's surface, but as a result of the discovery of new data that forced most of the sceptics to accept the fact of continental drift re-evaluation and the mechanisms that had already been proposed. In the wake of the Second World War, oceanographic surveys began to turn up additional observations that supported continental drift theory. Evidence from the study of the orientations of magnetic minerals in rocks showed that either the Earth's magnetic poles had shifted position over time or the continents had shifted positions over time. Depth soundings throughout the oceans revealed ranges of submerged mountains in all major ocean basins that were connected into a single worldwide network. In addition to these mid-ocean ridges, a system of very deep ocean trenches also existed, often in positions adjacent to large volcanic islands (e.g. Japan) or tall mountain ranges that included volcanoes on the continents (e.g. The Andes). Most convincing,(1) the rocks of the sea floor exhibited a regular pattern of zones of normal magnetic polarity and reversed magnetic polarity that were arranged symmetrically on either side of the mid-ocean ridge system and (2) both volcanic and earthquake activity took place predominantly on the mid-ocean ridges and in the deep-ocean trenches. Based on these data it was clear that new ocean crust was being generated at the mid-ocean ridges, spreading out away from these ridges over time, and then disappearing down the deep-ocean trenches. In the middle and late 1950s these observations were collected together to form an update to Wegener's theory of *Kontinentalverschiebung*, which received a new name: the theory of sea-floor spreading, ironically harking back to Holmes' mechanism first proposed over a quarter of a century before.

Subsequent geophysical studies of the sea floor and deep mantle, along with theoretical studies of the dynamics of the Earth's interior, have updated and refined Holmes' original scenario, which now goes by the name 'plate tectonics'. This term refers to the collection of processes and phenomena related to the fact that the Earth's surface has been mobile over geological timescales with the crust of continental landmasses drifting on the surface of a sea of mantle organized into slowly convecting cells by heat flow from the Earth's core. Because of their lower

RIGHT Diagram of major plate tectonics provinces. See text for discussion.

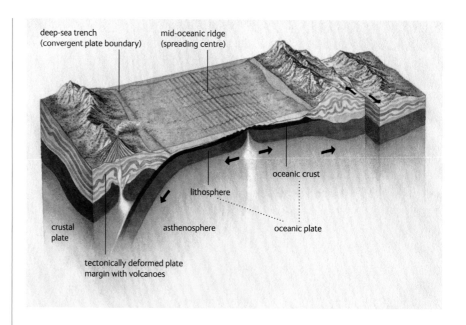

deep-sea trench
(convergent plate boundary)

mid-oceanic ridge
(spreading centre)

oceanic crust

lithosphere

crustal
plate

asthenosphere

oceanic plate

tectonically deformed plate
margin with volcanoes

density, the stable and relatively rigid continents ride passively, as plates, on top of heavier, denser oceanic crust that is created by volcanic eruptions at divergent plate boundaries (= the mid-ocean ridges, see above) and destroyed at convergent plate boundaries (= the deep-sea trenches). These convergent plate boundaries are themselves often sites of unusually frequent and intense earthquakes and volcanoes as ocean crust, along with the upper mantle layers, descends into the deeper mantle and any entrapped fragments of continental crust with associated low-density sediments are melted and rise back to the Earth's surface as magma.

After many years of being dismissed as part of the geological 'lunatic fringe', the speculations advanced by Wegener, his predecessors, his colleagues and his intellectual descendants have been proved correct after all. But, specification of the theory of plate tectonics, like specification of Darwin's theory of natural selection, was only the end of the beginning of the story. Once plate tectonics had been accepted by the scientific community, essentially all previous geographic observations made by geologists and palaeontologists over the preceding century had to be reinterpreted in light of this new theory in order to obtain a detailed understanding of Earth history. This process continues to the present day, with much contemporary research effort being expended on mapping ancient continental positions and the shapes of ocean basins in the distant past and using the palaeogeographic maps that result to inform investigations ranging from discovering patterns of species migration to charting the effect of climate change in the distant past. At present these efforts have produced an understanding of Earth history sufficient to allow palaeontologists to accurately locate the position of virtually any contemporary location on the Earth surface to its position in the geological past to an average accuracy of about ± 10° of longitude and latitude for locations on the stable regions of the major continental plates.

STRATIGRAPHY AND THE GEOLOGICAL TIMESCALE

In order to study extinctions in the fossil record it is necessary to determine the relative timing of species' appearances and disappearances. This is done via reference to the geological timescale, which is the system geologists have developed to summarize the sequence of geological events in time. The main principle by which the geological timescale was devised was first established by the Danish Catholic Bishop Niels Stensen, also called Nicolas Steno. In his studies of the rocks in Tuscany, Stensen reasoned that, because the rock layers at the bottom of a sequence must have existed before the overlying rock layers could be deposited, these basal rock layers must be older than those lying above them. Applying this 'Principle of Superposition' to a hypothetical sequence (see below) the ordering of rock types in age is the reverse of their order in distance from the sequence's base, with the youngest rock layers residing at the top of the pile and the oldest at the bottom. By the same reasoning fossils that occur in rock layers at the bottom of a sedimentary rock sequence must be older than those that occur in rock layers at the sequence's top.

Fossils do not often occur continuously throughout layers of sedimentary rock. In most sequences, finding a fossil is a comparatively rare – and so a special – event. Palaeontologists search sedimentary rock bodies for fossils and, when they find one, they usually collect it so that it can be identified by comparing it with other fossils in a museum collection. Regardless of whether the fossil can be identified at the outcrop, the palaeontologist carefully notes its position in the rock sequence, usually as a distance from the base or the top of the outcrop or core. This is the fossil's stratigraphic location or distance.

As more discoveries are made a picture begins to emerge of how each fossil species is distributed in the rock body (see p.26 top left), how low and how high in the sequence the fossil species' occurrences reach, how frequently the fossil occurs, and how its occurrence pattern compares with those of other species in the same and different sedimentary rock sequence(s).

Older

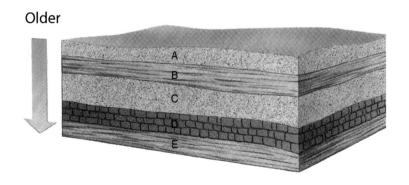

LEFT Steno's Principle of Superposition which is used to infer the relative ages of layers of sedimentary rocks. In this diagram layer E (bottom) is the oldest and layer A (top) the youngest.

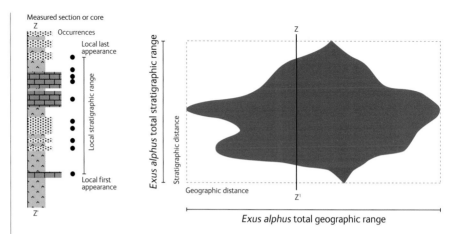

RIGHT Two diagrams that illustrate the concept of the fossil biozone. Left, concept of the biozone as inferred from occurrences (circles) in a section or core of sedimentary rock layers. Right, the total species biozone which (theoretically) is composed of information from all geographic regions and all sections/cores containing the species. Any particular section or core represents only a single part of the complexly structured region of space and time that marks out the total zone of species occurrence.

Researchers know from the study of modern species that some are restricted to particular environments while others range across many environments. For example, the remains of a lacustrine fish species will only be found in lake sediments, whereas the pollen from a pine tree growing on the bank of the lake could, in principle, be found over a wide range of different environments including the adjacent lake sediments. Since these different environments are characterized by different types of sediments (which eventually turn into different sedimentary rock types), some fossils are confined to particular types of sedimentary rocks whereas others can be found in many different rock types, each of which reflects a different environment. Those organismal groups that have a high probability of becoming fossils and a wide environmental tolerance are particularly important for establishing time relations among sequences of sedimentary rocks. Geologists call these index fossils.

Index fossils are used for matching up the time sequences of rock layers that exist at different locations on the Earth's surface. You can think of a fossil species' distribution in time and space as forming an irregular zone in the rocks (see above right). Each species' zone has a low point, which represents the species' true origination, and a high point, which represents its true extinction. The shape of the zone is determined by the species' patterns of geographic expansion and contraction, which reflect varying times of appearance (migration into) and disappearance (extirpation in) local environments.

Index fossils are species whose zones – also termed biozones – are geographically widespread but temporally short, so that they mark off relatively small intervals of geological time. Ideally index fossils will also be characterized by rapid migration outward from their points of origin, followed by their common and ubiquitous occurrence across many different environments, and ending with their relatively rapid extinction at all the localities in which they are found. While no fossil species conforms strictly to this ideal index fossil concept, the distributions of some can be treated as if they effectively mark successive zones of time in the fossil record. These biozones are used as one of the primary building blocks of the geological timescale.

Let's take a Late Cretaceous time interval as an example. The uppermost three biozones of the Cretaceous time interval are show in Table 2.

Table 2. Stratigraphic classification of the uppermost Cretaceous time interval.

Biozone	Age/Stage	Epoch/ Series	Period/ System	Era/ Erathem	Eon/ Eonathem
Anapachydiscus terminus Biozone					
Anapachydiscus fresvillensis Biozone	Maastrichtian	Upper	Cretaceous	Mesozoic	Phanerozoic
Pachydiscus neubergicus Biozone					

All of the layers of sedimentary rock that were deposited during the time interval when the ammonite species *Anapachydiscus terminus* existed, fall within the *Anapachydiscus terminus* biozone. This biozone is the uppermost unit of a larger, more inclusive interval of geological time known as the Maastrichtian. Geologists refer to all rocks deposited during this larger time interval as belonging to the Maastrichtian Age. The stratigraphic (= rock) interval encompassed by rocks assigned to the Maastrichtian Age is referred to as the Maastrichtian Stage.

The Maastrichtian Age/Stage, in turn, is the uppermost age/stage of a more inclusive time interval, Upper Cretaceous Epoch/Series, which, in turn, is the uppermost epoch/series of the Cretaceous Period/System which, in turn, is the uppermost period/system of the Mesozoic Era/Erathem. The Mesozoic Era/Erathem is a constituent interval of the Phanerozoic Eon/Eonathem.

Biozones are the primary means of determining time throughout the Phanerozoic, which comprises the last 540 million years of Earth history. Prior to this the diversity of life appears to have been quite low, characterized by organisms at the sub-eukaryote (naked DNA/RNA; Archean Eon/Eonathem) or eukaryote grade of organization (DNA/RNA enclosed in a lipid membrane; Proterozoic Eon/ Eonathem). All rocks existing on the Earth, and so the entire history of the planet, can be fit into the time intervals that together comprise the geological timescale. Like the periodic table of the elements, the geological timescale represents a basic tool of science. It is a summary of the work of countless geologists and palaeontologists, who have contributed primary stratigraphic observations to it for over 200 years. A complete listing of the current geological timescale is shown on p.28.

It is important to remember that the geological timescale is a relative timescale. Owing to the logic of the Principle of Superposition, the timescale exists, is accurate and has proved to be very useful irrespective of whether absolute dates are assigned to the time interval boundaries. In some cases – mostly when layers of volcanic sediments that contain radioactive minerals have been deposited at or near a time-interval boundary – it is possible to estimate the absolute age of a timescale boundary using radioisotopic methods. Most of the major boundaries have been assigned absolute ages in this manner. However, many of these absolute age

Geological timescale — Phanerozoic (Mesozoic–Caenozoic)

Eonathem/Eon	Erathem/Era	System/Period	Series/Epoch	Stage/Age	Age
Phanerozoic	Mesozoic	Cretaceous	Lower	Berriasian	145.0
				Valanginian	139.8
				Hauterivian	132.9
				Barremian	129.4
				Aptian	125.0
				Albian	113.0
			Upper	Cenomanian	100.5
				Turonian	93.9
				Coniacian	89.8
				Santonian	86.3
				Campanian	83.6
				Maastrichtian	72.1
	Caenozoic	Palaeogene	Palaeocene	Danian	66.0
				Selandian	61.6
				Thanetian	59.2
			Eocene	Ypresian	56.0
				Lutetian	47.8
				Bartonian	41.3
				Priabonian	38.0
			Oligocene	Rupelian	33.9
				Chattian	28.1
		Neogene	Miocene	Aquitanian	23.03
				Burdigalian	20.44
				Langhian	15.97
				Serravallian	13.82
				Tortonian	11.62
				Messinian	7.246
			Pliocene	Zanclean	5.333
				Piacenzian	3.600
		Quaternary	Pleistocene	Gelasian	2.588
				Calabrian	1.806
				Middle	0.781
				Upper	0.126
			Holocene		0.0117

Geological timescale — Phanerozoic (Palaeozoic–Mesozoic)

Eonathem/Eon	Erathem/Era	System/Period	Series/Epoch	Stage/Age	Age
Phanerozoic	Palaeozoic	Carboniferous	Mississippian (Lower)	Tournaisian	358.9
			Mississippian (Middle)	Visean	346.7
			Mississippian (Upper)	Serpukhovian	330.9
			Pennsylvanian (Lower)	Bashkirian	323.2
			Pennsylvanian (Middle)	Moscovian	315.2
			Pennsylvanian (Upper)	Kasimovian	307.0
				Gzhelian	303.7
		Permian	Cisuralian	Asselian	298.9
				Sakmarian	295.5
				Artinskian	290.1
				Kungurian	279.3
			Guadalupian	Roadian	272.3
				Wordian	268.8
				Capitanian	265.1
			Lopingian	Wuchiapingian	259.9
				Changhsingian	254.2
	Mesozoic	Triassic	Lower	Induan	252.2
				Olenekian	251.2
			Middle	Anisian	247.2
				Ladinian	242
			Upper	Carnian	235
				Norian	228
				Rhaetian	208.5
		Jurassic	Lower	Hettangian	201.3
				Sinemurian	199.3
				Pliensbachian	190.8
				Toarcian	182.7
			Middle	Aalenian	174.1
				Bajocian	170.3
				Bathonian	168.3
				Callovian	166.1
			Upper	Oxfordian	163.5
				Kimmeridgian	157.3
				Tithonian	152.1
					145.0

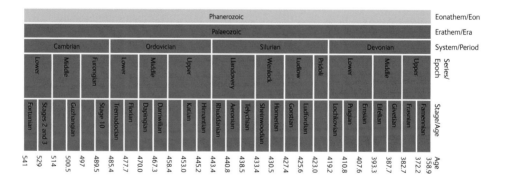

ABOVE The geological timescale for 2012 released by the International Commission on Stratigraphy. To get a sense for how much change the timescale undergoes over the course of just a few years compare this diagram with the previous timescale 2010 (http://www.stratigraphy.org/column.php?id=Chart/Time%20Scale) or the Geological Society of America (http://www.geosociety.org/science/timescale/). There are substantial changes in the boundary age dates, but much less so in the names and sequences of the major ages and epochs.

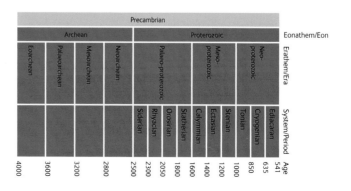

estimates remain subject to considerable error. While the age estimates of most age/stage boundaries have, on the whole, grown progressively more accurate over time, the dates assigned to all age/stage boundaries have changed and will continue to change for the foreseeable future as samples are recovered from new locations, as new samples are collected from old localities, and as the instruments that measure the abundance of radioisotopes and the daughter products of radioactive decay improve in sensitivity. But the relative sequence of time intervals has remained much more stable than the absolute dates assigned to their boundaries, remarkably so in the Mesozoic and Caenozoic Erathems. As we will see below, extinction events have often been used to define intervals of geological time, with individual species' extinctions being used to define many biozone boundaries, minor extinction events defining age/stage boundaries, and major extinction events defining system/series and era/erathem boundaries.

WHERE DO THE DATA FOR EXTINCTION STUDIES COME FROM?

Oddly enough, one of the things often left out of discussions about the patterns and prospective causes of ancient extinction events is an explanation of the sources of data on which various conclusions are based. Ultimately, almost all data of palaeontology – except in the case of purely theoretical studies – can (usually) be traced back to a lone palaeontologist sitting on a rock-strewn slope recording the position of a newly found fossil's geographic location and position in the sequence of layered sedimentary rocks before bagging it up for transportation to a collection of similar specimens in a museum, university geology department, or commercial laboratory (see top p.30). As noted above, the geographic record locates the fossil in space, not only the space of the modern world but also its location on the Earth's surface in ancient times when the continents, and fragments thereof, may have occupied very different positions. The depth record, coupled with the geographic record, locates the specimen in time according to the geological timescale. Often the information provided by other fossils found in the same general vicinity is also critical to determining each specimen's relative age.

Surprisingly few fossils are (or can be) identified accurately to the level of the species at the time they are collected in the field. In order to be identified with confidence most fossil specimens must be transported to places where they can be compared with drawings, photographs and descriptions of other specimens collected by other palaeontologists, or, better still, compared directly with other specimens in a museum collection that have been identified by specialists.

Palaeontologists who study ancient extinctions are primarily interested in a special subset of these data that have been collected at such great financial and (often) personal cost by generations of their colleagues. Extinction studies focus on the stratigraphically highest observed appearance of a species, a taxonomic

ABOVE Virtually all palaeontological data are collected by palaeontologists searching for fossils the field. Photograph of the author searching for fossils in rocks of Upper Carboniferous age in North-Central Texas, USA.

genus or a taxonomic family to which the specimen belongs. This is an unusual sort of observation, the interface between what has been observed and what has not been observed. Most sciences are concerned with documenting the characteristics of, and understanding the processes that cause, objects, specimens and species to come into existence or to be able to maintain their existence in a particular place at a particular time. But extinction data are different. In these cases we are interested in the negative side of the same equation - the reasons species disappear from the fossil record.

Scientists and philosophers call the data that come from failing to make an observation negative evidence or evidence of absence. We run into negative evidence all the time in our everyday lives. For example, when a doctor screens a patient for cancer they look at tissue samples to determine whether cancer cells are present. A favourable result medically is the one that fails to find evidence of cancer. A more common example occurs when you visit a new supermarket looking for your favourite brand of (say) breakfast cereal, but fail to find it. In both cases the evidence indicates that the objects being sought are not present. However, the data that were (not) collected do not allow you to distinguish between true absence and the mistaken appearance of absence that could result from you looking in the wrong places or not having access to the technology that would reveal the object's presence to you. In the latter example this ambiguity is annoying. In the former it is potentially fatal.

Extinction data are particularly problematic in this regard. As I noted above, finding a fossil is, in most instances, a relatively rare event. When a palaeontologist samples a layered rock sequence for fossils there are often considerable gaps between the horizons at which fossils of the same species are found. For example, the graph right the pattern of occurrence of ammonite species in the rock sequence leading up to the Cretaceous–Palaeogene (K–Pg) boundary on Seymour Island in Antarctica. In this diagram the level of the K–Pg boundary is shown by

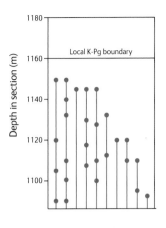

the horizontal line. If we accept the fossil record at face value, none of the ammonite species extend to the level (= time) of the boundary. The pattern of apparent extinction would suggest that all 10 ammonite species recorded from this stratigraphic section became extinct prior to the K–Pg boundary.

But the pattern of occurrence for each species is very sparse, with long gaps between the occurrences of specimens belonging to each species (represented by circles on the diagram) within their known stratigraphic ranges (represented by the vertical lines joining the occurrence record circles). Throughout the time interval represented by these gaps each fossil species must have existed even though there is no record of its existence on Seymour Island. Beyond the last occurrence in each species' fossil record, however, there is no evidence of any of these ammonite species existing anywhere else in the world ever again.

Given the inherent 'gappiness' in these species' fossil records, how do we know that any given species did not persist somewhere in the world beyond its last recorded appearance on Seymour Island? In other words, how far beyond these last appearance horizons do we need to go before we are confident each species really was extinct? Could a few of the species have persisted to the level of the K–Pg boundary and become extinct at that time? Indeed, could a few have persisted beyond the K–Pg boundary and into the Palaeogene Age? These are all questions palaeontologists who study ancient extinctions ponder and try to find answers for.

Unfortunately, the stratigraphic palaeontological literature is too vast for any researcher to be able to create a dataset that authoritatively summarizes the extinction horizons for all 500,000 or so fossil species that have been described. To summarize the general trends that exist in these species-level data, synoptic datasets of more tractable size have been compiled by specialists at the levels of the taxonomic genus and family. Starting with compendia of taxonomic and age data such as *The Fossil Record* (Harland *et al.*, 1967) the *Treatise of Invertebrate Palaeontology* (1953 to present) and *The Fossil Record 2* (Benton 1993) summaries of around 35,000 fossil genera and 4,000 fossil families have been assembled and published by data compilers – notably the late University of Chicago palaeontologist J. J. (Jack) Sepkoski Jr – and checked, amended and rechecked by an army of specialists in various taxonomic groups. This work is continued on a voluntary basis as the core of the Paleobiology Database Project (http://paleodb.org/cgi-bin/bridge.pl).

Taxonomic compendia aggregate species-level data and treat higher-level taxonomic categories as if they had the same ontological status as species. However, most evolutionary biologists would argue there are fundamental differences between the concepts of a species and a genus or family. The former represents a level of biological organization that is the product of natural processes. The latter represent the product of a human attempt to organize biodiversity data into a smaller number of hierarchical categories than is possible under, or logically consistent with, Darwinian evolution. Irrespective of the differences between species and higher taxonomic categories, such summaries are useful in revealing general patterns in the fossil record. Genus-level and family-level data can, have, and do serve as proxies for species-level data. They are especially useful for indicating where further species-level research needs to be done and in testing certain kinds of hypotheses.

In addition to aggregating information across sets of species, these taxonomic compendia also aggregate the stratigraphic information included in species descriptions into higher stratigraphic categories, usually at the level of the stratigraphic age though in some cases subage units are used. This is an eminently pragmatic convention as the stratigraphic levels recorded in the species-level literature (usually biozones) are complex to interpret under the best of circumstances, much less to summarize. However, use of aggregated stratigraphic data results in the binning of stratigraphic information into relatively coarse time intervals.

For example, in most palaeontological compendia all extinctions occurring at any point in the Maastrichtian Age are listed as 'Maastrichtian'. While this label should be read as 'extinction sometime in the Maastrichtian' many data analysts have interpreted the results of analyses of taxonomic compendia data as if all Maastrichtian extinctions occurred precisely at the K–Pg boundary. This practice artificially extends the stratigraphic range of at least one and possibly all species assigned to a taxonomic group to the upper limit of the stratigraphic interval despite the fact that no member of the group may ever have been observed at that horizon (see p.31); the palaeontological equivalent of 'rounding off' the stratigraphic data to the nearest inclusive age. While this practice greatly simplifies the representation of time in palaeontological datasets, the gain in palaeontologist's ability to assess general trends in data derived from myriad collections of particular specimens in the field is purchased at the price of temporal accuracy. Since evaluation of the conformance of different data to the predictions of different causal scenarios requires the accuracy of species occurrences in time to be known, the dangers of assuming - rather than observing - any particular extinction geometry are obvious. As an alternative to the use of taxonomic compendia, palaeontologists may opt to work only with high-resolution data derived directly from the observation of specific species occurrences in the field. Data for both sorts of studies exist for some groups, but not for the majority of groups across the major stratigraphic extinction horizons.

Irrespective of these complications, stratigraphically binned data have played a crucial role in revealing major patterns of variation in the history of extinctions on Earth and do provide a valid means of testing certain hypotheses. Such data must be treated with caution and should not be over-interpreted. But they represent the best summaries palaeontologists currently have to aid our understanding of many aspects of the fossil record.

4 Kinds of extinction

EXTINCTION DATA AND TRENDS

The first diagram of the biodiversity or, to use the more technically correct term, the richness of life through geological time was published by John Phillips in 1860 (see right). Phillips' diagram shows this history as having three phases in the form of diversity radiations, each of which is separated by distinct intervals during which richness fell, one at the close of the Palaeozoic era and one at the close of the Mesozoic era. A smaller and less well-defined drop in richness was also proposed to characterize the middle of the Palaeozoic. In addition, this diagram shows that richness increased in each of the major geological time intervals reaching a peak in the modern world.

Though Phillips' biodiversity diagram was used in his own day to advocate acceptance of the Palaeozoic, Mesozoic and Cænozoic (= Caenozoic) as natural divisions of geological time, it took more than a century for palaeontologists to become seriously interested in the study of the patterns evident in his diagram. This delay reflects the fact that many late nineteenth and early twentieth century geologists regarded Phillips' diagram as being based on too little data to offer a definitive summary of life's rich history.

Palaeontologists turned their attention back to diagrams such as Phillips' in the 1960s and 1970s after comprehensive summaries of stratigraphic data for large numbers of fossil families and genera began to be published. This rejuvenation of extinction studies was begun by Otto Schindewolf in the 1950s, but taken forward actively by Norman Newell who, throughout the 1960s, argued that biotic crises – what we have come to call 'mass extinctions' – were real phenomena of the fossil record and not, as Darwin had supposed, examples of the imperfections of that record. In supporting his case Newell assembled the first comprehensive datasets or compendia of extinction patterns across many fossil groups (see below).

OPPOSITE A death assemblage of marine invertebrate species brought together by wave action along the strand line of a beach. Note the preponderance of bivalves and the fact that the two valves of the bivalve shells are virtually completely disarticulated. This disarticulation only happens after the bivalve dies and the ligament that holds two valves together in life rots away.

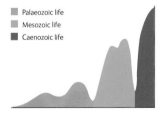

■ Palaeozoic life
■ Mesozoic life
■ Caenozoic life

ABOVE John Phillips' 1860 diagram of Phanerozoic biodiversity.

LEFT Norman Newell's 1963 summary of Phanerozoic biodiversity for marine invertebrate families.

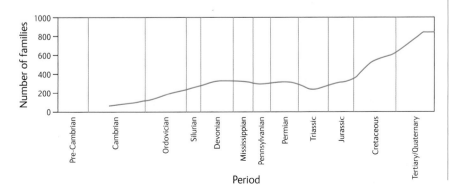

RIGHT Two of J. J. (Jack) Sepkoski's
summaries of the biodiversity
history of marine animals based
on family-level (above) and genus
level (below) data. V., Vendian;
Camb., Cambrian; Ord., Ordovician;
S., Silurian; Dev., Devonian; Carb.,
Carboniferous; Pr., Permian;
Tr., Triassic; Jur., Jurassic; Cret.,
Cretaceous; Palaeog., Palaeogene.
(Redrawn from Sepkoski 1997.)

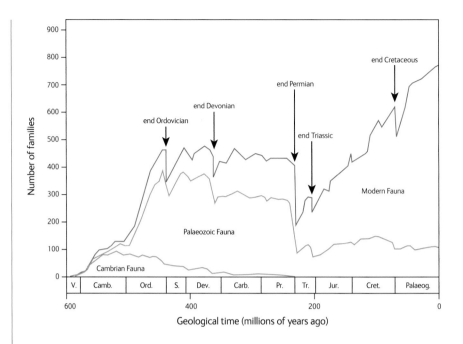

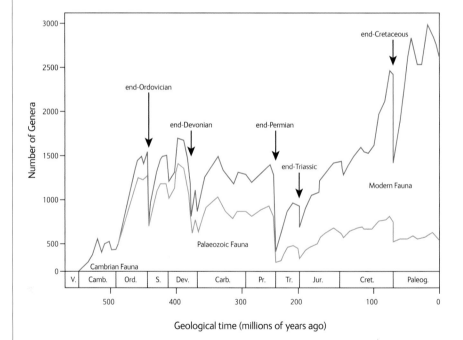

RIGHT Two of J. J. (Jack) Sepkoski's summaries of the biodiversity history of marine animals based on family-level (above) and genus level (below) data. V., Vendian; Camb., Cambrian; Ord., Ordovician; S., Silurian; Dev., Devonian; Carb., Carboniferous; Pr., Permian; Tr., Triassic; Jur., Jurassic; Cret., Cretaceous; Palaeog., Palaeogene. (Redrawn from Sepkoski 1997.)

In the wake of Newell's publications, a number of taxonomic compendia were produced (e.g. Harland *et al.*, 1967, Van Valen 1973) and used to explore various long-term trends in the fossil record. But no single compiler of such data was more assiduous than the late J. J. (Jack) Sepkoski Jr. In 1981 Sepkoski published a summary of the stratigraphic ranges of 2,800 families of marine invertebrates, the data from which he assembled into a diagram that constituted an update of Phillips' 1860 summary (Sepkoski 1981, see top).

While the level of stratigraphic resolution, number of taxa employed, level of documentation, and overall quantitative treatment of the data had improved greatly in the almost 150 years since Phillips' original work, the same general patterns are present in Newell's and Sepkoski's diagrams including the diversity reductions at the end of the Palaeozoic and Mesozoic, fluctuations in the Middle Palaeozoic, and the overall increase in diversity throughout the Phanerozoic. Sepkoski continued to collect data from the palaeontological literature after 1982, revising the family-level curve and releasing a genus-level dataset in the middle 1980s (see p.36 bottom). In both cases the general patterns summarized in the original family-level curve were also present in the expanded family and genus-level taxonomic diversity data.

The Phanerozoic diversity curves assembled by Sepkoski and others reflect the complex interplay between the evolutionary processes of the origination (= speciation) and the extinction of taxonomic groups. Since the databases used to construct these curves record the appearances and disappearances of individual groups it is possible to tease apart these two aspects of biodiversity control.

There are four extinction/origination indices palaeontologists have used to describe fossil biodiversity/richness data. Since both the form of these indices, and the issues associated with each, are the same for origination and extinction data only the extinction indices will be given here.

$$\text{Total extinction} = \text{no. of taxa becoming extinct during interval}$$

$$\text{Proportional extinction} = \frac{\text{total extinction}}{\text{no. of taxa existing during interval}}$$

$$\text{Time-normalized extinction} = \frac{\text{total extinction}}{\text{duration of time interval}}$$

$$\text{Per-taxon extinction} = \frac{\text{proportional extinction}}{\text{duration of time interval}}$$

The total extinction index is the simplest to calculate, but is biased by a number of factors. The proportional extinction index accounts for the number of taxa at risk of extinction in any time interval, but not of the size of the time interval. The time-normalized index takes the duration of the interval over which extinction-related processes operate into consideration, but not the number of taxa at risk of extinction. The per-taxon index is a probabilistic measure of extinction intensity that takes the number of taxa at risk and the duration of the time interval into account, but returns an extinction intensity estimate that is biased by interval length under a wide range of extinction models (see Foote 1994). None of these indices is perfect, but all capture important aspects of extinction data.

Looking at the history of stage-level extinctions for marine invertebrate genera it is obvious that, whereas extinction susceptibility may be constant across taxonomic

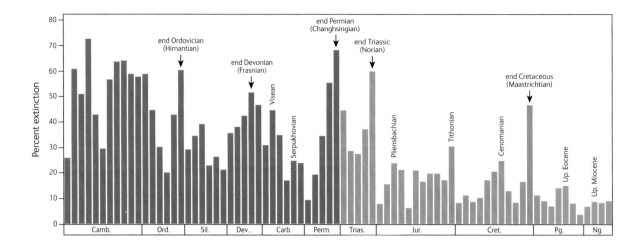

groups, extinction intensity has varied markedly throughout geological time (see
above). Intervals of relatively high extinction intensities are separated from one
another by intensity troughs. In some cases the intensity peaks stand well above the
local level of average extinction intensity (e.g. the Norian and Maastrichtian peaks).
In others the peak represents the culmination of a progressive build-up in extinction
intensity across several stratigraphic stages (e.g. the Frasnian and Cenomanian
peaks). In addition there is a pronounced decrease in the average level of extinction
intensity from older to younger time intervals.

MASS EXTINCTIONS

The concept of mass extinction has a complex history. Georges Cuvier formulated it
in the early 1800s after noting the juxtaposition of fossils of unusual horse-like and
dog-like creatures with seashells in several localities within the Paris Basin. Cuvier,
along with his collaborator Alexandre Brongniart, interpreted these associations as
indicating an environmental catastrophe had taken place in Earth's distant past, a
catastrophe that exhibited a scale and intensity unprecedented in human history.
This event, whatever its cause, affected terrestrial and shallow marine environments
together, sweeping up organisms that lived in both areas indiscriminately and
depositing them in assemblages that could not be considered natural. Cuvier referred
to these events as 'revolutions' drawing on the most obvious political parallel of
his own day: the French Revolution. Originally in a 1810 report on his Paris Basin
stratigraphic investigations, and from 1812 onwards in his monograph on fossil fish
(see Rudwick 1997), Cuvier expanded and developed this concept of revolutionary
extinctions that resulted in whole groups of organisms being annihilated suddenly
such that only a few relict survivors were left.

   British geologists largely rejected Cuvier's revolutions in the history of life,
characterizing this view of life's history as 'catastrophism'. Led by Charles Lyell,

BELOW Baron Georges Cuvier,
originator of the concept of mass
extinction.

and supported by Charles Darwin, they preferred to regard Earth history as being the product of the same processes they saw operating in the modern world – the doctrine of actualism (sometimes mistakenly referred to as uniformitarianism, see Gould 1987).

As we have seen, Darwin's theory of natural selection accommodated the idea of extinction, but Darwin was uncomfortable with the convenience of explaining the data of nature by invoking mysterious, catastrophic processes that caused large sets of ecologically disparate species to die off simultaneously. In part, Darwin rejected Cuvier's views out of a desire to make his ideas about the nature of evolution, and the process of natural selection, as scientifically rigorous, and as free from the need to invoke seemingly mystical explanations, as possible. Consequently, Darwin preferred to explain Cuvier's revolutions as disruptions in the continuity of evolutionary processes which were the result of imperfections of the geological record.

The outcome of what has come to be called the catastrophist–uniformitarian debates in the middle 1800s was that extinction in general, and mass extinction in particular, came to be regarded as non-issues in most palaeontological circles throughout the first half of the twentieth century. In particular, the stratigraphic records of the transition between the Palaeozoic and Mesozoic Eras, and between the Mesozoic and Caenozoic Eras, were found to be marked by substantial taxonomic and stratigraphic discontinuities. Of course, these are precisely the same intervals at which John Phillips' 1860 biodiversity curve showed substantial declines in species numbers (see p.35).

By the early 1960s, under the influence of Schindewolf, Newell and others, the idea of episodic biodiversity crises or 'mass extinctions' had (re)gained acceptance. The (now common) inference of taxonomic discontinuities across a few major stratigraphic horizons was rendered consistent with the (then) current understanding of evolution by arguing that, despite their existence, mass extinctions were caused by actualistic processes (e.g. climate change, sea-level fluctuation) and took place over millions of years. Regardless, as late as the 1970s there was little consensus among palaeontologists regarding which ancient extinctions were mass extinctions (other than those associated with Phillips' original end-Palaeozoic and end-Mesozoic transitions) and how long the process of mass extinction took.

Cuvier's original evidence for mass extinction was the juxtaposition of marine and terrestrial animals in the same fossil deposit. Although most fossils are first noticed as isolated occurrences, concentrations of fossil shells and/or bones do exist. In most cases these concentrations come together after the organisms have died, usually accumulating passively over long periods of time as a result of normal sedimentation processes. An example of this is the 'ooze' that blankets much of the deep-ocean floor. This deposit is created by the shells of microscopic plankton falling through the water column and accumulating over time on the seabed. In some places the ooze can be hundreds of metres thick and include the shells of species that became extinct tens of millions of years ago.

In other cases the processes responsible for the accumulation of dead organisms are active, such as waves and currents along a beach, a flooded riverbed, a tsunami, a landslide or a density current on the continental slope. These natural processes sweep objects up over a wide area, carry them suspended in the current for long distances, and then set them down when current velocity slows. As a by-product of this process the current sorts the objects it carries into groups with similar hydrodynamic properties.

Both these types of accumulations are termed 'death assemblages' by palaeoecologists, since the organic remains involved all belong to long-dead organisms that have been removed from the contexts in which they once lived. Death assemblage deposits are by far the most common in the fossil record. However, there is another category of natural assemblage preservation that is much rarer, but far more interesting from a biological point of view. This is the so-called 'life assemblage'.

Life assemblages occur when an individual or group of organisms is overcome suddenly by some natural process (e.g. flood, debris flow, sandstorm, dust-storm, volcanic ash fall) that causes the individual or group to be buried rapidly and completely and then to be fossilized (see opposite). Such fossil assemblages are special because the individuals themselves are preserved in as near to their living states as possible and because a collection of individuals often preserves important aspects of the spatial systems, ecological systems, behavioural systems, developmental systems and in some cases even the social systems of which the living organisms were part. In virtually all cases burial must be very rapid and complete for this information to be preserved. Certainly from the point of view of the organisms involved, the process that led to their preservation would have been sudden, catastrophic and, of course, fatal. As a result, some examples of these life assemblages, especially when the subjects are vertebrates, have been termed mass-mortality sites, or mass-kill deposits the latter of which are analogous, at least in part, to sites of the mass killing of modern animals by certain social species (e.g. sharks, whales) as well as the mass-killing fields that litter human social history.

With concepts such as life assemblages of fossils being described with emotive terms such as 'mass-kill' deposits, and with concrete examples of the sudden extermination of large numbers of animals by human hunting parties, it is little wonder that the term 'mass extinction' began to be used to describe aggregations of fossils, first by popularisers of palaeontology and then by palaeontologists themselves. The first use of the term I have come across was to describe the Pleistocene megafaunal extinctions to which human hunters may indeed have contributed. Later though, the term mass extinction was used to refer to any – and seemingly all – instances of elevated extinction rates irrespective of the data used and/or of the possible causes of the extinction. In my personal view, appropriation of this term from the ecological and taphonomic literature has been the source of much confusion because it implies a sudden, catastrophic and externally forced cause when, in reality, the data being described may not support such an interpretation.

Lettre I.

LEFT AND BELOW Examples of specimens from the *Iguanodon* mass-mortality site at Bernissart, Belgium. Left, drawing of a specimen in situ. Below, reconstructed *Iguanodon* specimens as they might have appeared in life based on the information contained in this extraordinarily well-preserved and complete association of articulated fossil remains.

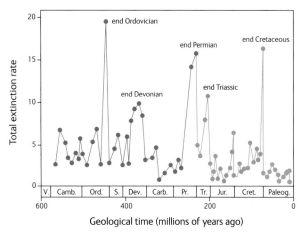

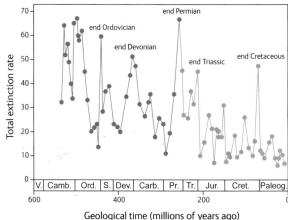

Plots of stage-level percent extinction data for marine families (left, after Raup and Sepkoski 1982) and genera (right, after Raup and Sepkoski 1986).

BELOW Stage-level Phanerozoic extinction history for marine genera ordered by percent extinction intensity. In many cases peaks smaller than the Big Five extinction events (see text) have been referred to as representing 'mass extinctions' including time intervals that contain some of the lowest extinction-intensity levels e.g. Pleistocene. (Blue, Palaeozoic stages; green, Mesozoic stages; red, Caenozoic stages.)

In 1982 David Raup and Jack Sepkoski published a summary of Sepkoski's family-level extinction data and concluded that the magnitude of extinction that occurred during these intervals of geological time were sufficiently high to warrant designation formally as 'mass extinctions'. Shortly thereafter palaeontologists began to speak of the 'Big Five' mass extinctions. This report was followed in 1986 by a summary of genus-level data that Raup and Sepkoski presented as evidence confirming their earlier interpretation (see top left). In both cases it was noted by others that the degree of variation in the extinction data as a whole precluded definitive identification of these five events as representing a unified class of large events based on extinction magnitude data alone.

This argument can be appreciated simply by rearranging the histogram on p.38 so that the stages are ordered by extinction magnitude rather than by time (see top right). When plotted in this way the extinction-intensity data appear continuous with no discontinuity separating the Big Five mass-extinction events from the others. Indeed, these five extinction events do not even constitute a contiguous interval at the high end of the extinction-intensity spectrum, but are separated from one another by a series of equally large Early Palaeozoic events.

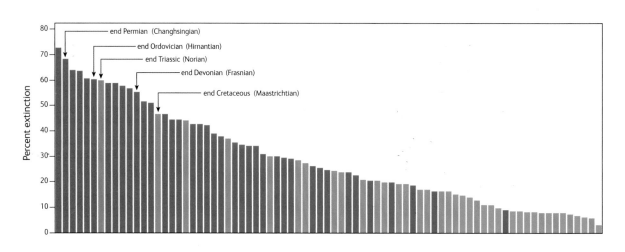

This rank-order extinction intensity plot also provides support for other insights into the character of ancient extinctions. The continuity of the extinction-intensity distribution implies that the process(es) responsible for both large and small events may be similar, but that the intensity with which these processes have operated over geological time has varied markedly. Evidently, the pattern of extinction intensity observed through the last 540 million years of Earth history is driven not so much by intrinsic differences in the processes that cause species to become extinct as by historical differences in the waxing and waning of these processes across geological time. In short, the fossil record of taxonomic extinction suggests that history matters. The continuity of this spectrum also explains why there is no commonly accepted definition of the term 'mass extinction' other than as a vague generic reference to a 'very large' extinction or to (the wag's favourite) 'the extinction event I'm trying to impress you with at the moment'.

To be fair, subsequent investigators have looked for, and in some cases found, evidence that does appear to distinguish some of the Big Five events from others. For example, during several (but not all) of these events both marine and terrestrial biotas appear to have been affected more-or-less equally whereas in lower intensity events these two ecological realms are usually decoupled. Likewise, in several of the Big Five events tropical biotas appear to have suffered differentially intense losses. Certain taxonomic groups also seem to have been more extinction prone than others. This cohort of 'at risk' groups that suffer repeatedly during the Big Five extinction events include ammonoids, trilobites, graptolites and reef-builders (e.g. corals, bryozoans, calcareous algae, stromatoporoids). In other cases potentially important patterns of variation have been found to characterize one event, but comparable data have yet to be collected for others. For example, David Jablonski has shown that, for smaller extinction events, mollusc species characterized by planktonic larval stages experience lower rates of extinction than those whose larva quickly settle to the sea floor, but during the end-Cretaceous extinction event both these groups suffered comparable losses. Whether this pattern also characterizes other members of the Big Five extinction club remains to be demonstrated.

Obviously much more research involving the collection of comparable data from different extinction events remains to be done. These comparisons are of paramount importance because, as David Raup has noted:

> There is no way of assessing cause and effect [in historical data] except to look for patterns of coincidence – and this requires multiple examinations of each cause-and-effect pair. If all extinction events are different the deciphering of any one of them will be next to impossible.
>
> DAVID M. RAUP (1991, p.151)

This is an excellent rule of thumb, not only for extinction studies but for any field in which hypotheses require evaluation by historical – as opposed to experimental – testing. As a matter of fact, this same principle stands at the heart of the logic that is applied to all scientific investigations – to all of science. All scientists look for patterns of coincidence in their data in order to establish the link between a proposed cause and its effect. In a laboratory experiment if it were the case that every time we applied the same change to a system a different effect was produced researchers could never understand, or predict, the behaviour of the system. This process is rendered more complex when historical data are the only data researchers have access to, but the logical principles remain the same.

Turning back to the traditional 'Big Five' extinction events, how large are these events really? Remember, the extinction intensities shown on p.38 refer to the percentages of marine invertebrate genera lost at some point during each stratigraphic age as a proportion of the total number of genera known to exist during that age. While species-level data have gone into estimating the generic stratigraphic ranges summarized by this diagram, this summary has been the product of over 100 years' patient work by generations of palaeontologists and biostratigraphers. It would be impossibly time consuming to go back through all the publications containing palaeontological data and tabulate these data for all fossil species. By making some assumptions about these family and/or generic data, however, we can estimate how many species extinctions these percentages imply.

Imagine that each species contains exactly the same number of individuals, say 10. Then, for the sake of argument, imagine all genera contain exactly the same number of species, also 10. Now imagine that all the families into which all the genera are placed also contain the same number of genera, 10 again. If we carry this thought experiment to its conclusion we will have created a model of taxonomic biodiversity that contains 1 million hypothetical individuals. It isn't a particularly realistic model since we know that there are many more groups composed of small numbers of taxa than of large numbers. But it's a simple model and we can always change it later.

Now imagine an extinction event that eliminates individuals randomly. Because we know how many individuals are assigned to each species, each genus, each family and so forth, and because we know that, for this hypothetical extinction event, there is no selectivity among individuals, we can calculate how many species would need to become extinct in order to result in the loss of any particular proportion of genera, or families, or any higher taxonomic group. So, for this model we know that if (say) 75% of the individuals are lost, none of the kingdoms and phyla will become extinct on average, 1 class will be eliminated, 10 orders, 14 families, 38 genera and some 70 or so species. These numbers might sound low. But remember, if only a single individual of a single species survives the random extinction event, the genus, family, class, order, phylum and kingdom to which that individual belongs will also survive.

Once we have this model expressed mathematically we can do the calculations any way we like, including derivation of an estimate of the number of species that need to be eliminated by extinction events in order to remove any given proportion of genera, or families. We now have a way of estimating the species extinctions the various stage-level percentages shown above imply. Not only this, we can run the experiment for as many different starting conditions as we like to make our estimates more realistic.

When this is done for the 'Big Five' extinction events the estimates listed in Table 3 are obtained.

Table 3. Estimates of species-level extinction intensities represented by family and genus data for the 'Big Five' ancient extinction events (data from Jablonski 1955).

Extinction	Age/Stage	Family data		Genus data	
		Observed %	Est. species loss (%)	Observed %	Est. species loss (%)
End-Ordovician	Hirnantian	26	77–91	60	82–88
Devonian	Frasnian	22	70–88	57	79–87
End-Permian	Changhsingian	51	93–97	82	93–97
End-Triassic	Norian	22	70–88	53	76–84
End-Cretaceous	Maastrichtian	16	57–83	47	71–81

Even acknowledging the fact that marine invertebrate species are not the sorts of species most people are familiar with and that the loss of even a large number of such species could be regarded as inconsequential to the hypothetical man-in-the-street, these are unquestionably large numbers. In fact, these results indicate such large levels of species loss in the case of the largest stage-level extinctions (e.g. the end-Permian and end-Ordovician events) that it might be difficult to understand why all life did not cease to exist. Quite possibly the reason is because these extinctions did not happen at the same time, but were spread out over the millions of years included in the time intervals assigned to these stages: 2.8 million years in the case of the Changhsingian, 1 million years in the case of the Hirnantian. Nevertheless, our simple model suggests the magnitudes of accumulated species loss that took these extinction events are nothing short of astonishing.

BACKGROUND EXTINCTIONS

Like so much in life, it's the big, unusual, sudden, destructive events that get all the attention, especially from the media. But this does not mean all extinctions other than the Big Five are uninteresting or unimportant.

In order to distinguish these lesser, stage-level extinction events from the Big Five 'mass' extinctions, Raup and Sepkoski coined the term 'background extinctions' (see p.46 top). The mass extinctions are seen by most researchers as unusual events that require unusual explanation. By inference then, the set of background extinctions are regarded as including extinctions caused by the normal Darwinian evolutionary processes of competition and natural selection. In terms of the raw numbers though, the background extinction category is by far the more important. Estimates suggest that over 95% of all species extinctions that have occurred in the history of life have taken place during the background extinction intervals. This fact alone makes the background extinction category an important, yet a curiously overlooked, aspect of the extinction record. But even more than this, as a set of observations the background extinction data are noteworthy because they exhibit an unusual and

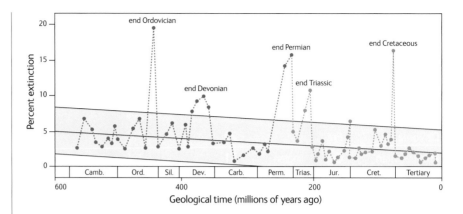

somewhat unexpected pattern of their own: the background extinction gradient.

This gradient is usually portrayed as a linear ramp with a negative slope through time (see above). Most commonly this ramp is described using trend line (linear regression) analysis, sometimes of the entire extinction-intensity dataset, sometimes with the Big Five mass extinctions removed. Results of this analysis indicate that the slope of the background extinction is statistically significant. Some researchers have suggested this phenomenon is an artefact of the fossil record and the geological timescale while others have interpreted it as a feature of the history of life.

In the case of the former interpretation, our ability to study modern species in greater detail than fossil species means representatives of many fossil genera and families have been discovered in the modern biota, often at some remove from the last apparent occurrence of fossil representatives of the same group. Since young fossil groups, on average, have a better likelihood of including recent species than older fossil groups, this fact biases the survivorship potential of young groups with a concomitant decrease in their apparent extinction potential relative to the same estimates calculated for older groups for which the existence of modern representatives is far less likely. In other words the average extinction rate is expected to decrease over time simply because we have a better sample of present biodiversity than we do of past biodiversity. This phenomenon has been termed the 'Pull of the Recent'. In addition, the average duration of a stratigraphic age is longer the further back in time we go. As a result, the number of species (and so genus, family, etc.) extinctions that can occur in each stage, on average, will be higher the further back in time we go. This phenomenon also acts to bias the extinction record towards higher extinction intensities in very ancient intervals. Both of these sources of bias exist in our extinction data. But is this the whole story in terms of the background extinction gradient?

The variation of extinction-intensity values about the linear trendline is such that the curve may be subdivided into segments. There is no significant trend to the background extinction-intensity data older than 350 million years ago (see opposite). Prior to this the decreasing linear trend is present, but variation of the data points about the regression line is stepped with high variation characterizing

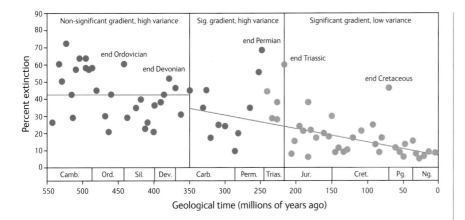

LEFT Stage-level Phanerozoic background extinction intensity data for marine invertebrate genera with the history subdivided into intervals based on best-fit linear-regression trend line analysis and assessment of patterns of variation about the trend line. Data points are colour coded by stratigraphic interval: red, Palaeozoic; blue, Mesozoic; green, Caenozoic. See text for discussion. (Redrawn from MacLeod 2004).

the interval from 200 to 350 million years ago and markedly lower levels of variation characterizing the interval from 0 to 200 million years ago. Neither of these patterns in the background extinction data is predicted from the Pull of the Recent or by the secular increase in the duration of stratigraphic ages. Moreover, a number of important environmental proxies (e.g. marine circulation, oxygenation, stability) that would be expected to influence marine and terrestrial biotas also exhibit long-term linear patterns through the Phanerozoic (Martin 1996, MacLeod 2003). While the long-term sources of bias in extinction data need to be appreciated and taken seriously, so do the data that show the Earth's environment has undergone a long-term pattern of development of its own, many aspects of which have been driven primarily by the diversification of life itself.

In addition to the factors discussed above three other hypotheses have been advanced to account for this feature of life's extinction history.

- A trend towards the adaptation of species to marginal environments with species that can survive in these habitats being inherently resistant to extinction pressures caused by environmental changes.

- An apparent trend of increase in the number of species per lineage over time due to a greater outcrop area of younger ages that is available for palaeontologists to search, which confers extinction resistance on younger lineages as a result of an apparently broader geographic distribution.

- Macroevolutionary effects of the invasion of new habitats by ecologically critical groups through the development of key adaptations.

Most likely all of these factors have played a role in the development and maintenance of the background extinction gradient over geological time. More research will be needed before the direct affects of each factor can be unravelled and understood in their proper context.

5 Causes of extinction

A BEWILDERING VARIETY OF MECHANISMS have been proposed to account for ancient extinctions. These have ranged from the obvious and testable (e.g. sea-level change, global cooling) through to the exotic and speculative (e.g. acid rain, global wildfires), and on to the untestable and, in some cases, ridiculous (e.g. cosmic radiation, hunting by aliens). I will not attempt to review all possible extinction mechanisms (see Benton 1990 for a review of dinosaur extinction theories), but rather will focus here on the subset of mechanisms contemporary palaeontologists and evolutionary biologists most often cite as being associated with the great extinction events documented in the fossil record.

In coming to grips with the question of extinction causes, proximate mechanisms need to be distinguished from ultimate mechanisms. A proximate extinction mechanism is one that can explain the cause of species extinction, but which itself may be caused by a number of different processes acting either singly (the SC scenario) or in concert (the MIC scenario, see p.16). For example, global cooling is an example of a proximate mechanism. There is abundant evidence to indicate warm-adapted species on the land and in the oceans often have a difficult time coping with extended periods of cold temperatures. However, there are a number of ways global cooling can be induced.

An ultimate extinction mechanism is one that causes extinctions (mostly) by inducing the operation of proximate extinction mechanisms that extend the effects of its operation well beyond the occurrence region. For example, volcanic eruptions are well-understood ultimate causal mechanisms for global environmental change. By lofting particles and releasing large amounts of gases high into the Earth's atmosphere volcanic eruptions can cause the amount of solar radiation reflected by the Earth into space to increase, thus leading to (among other things) a reduction in the mean temperature of the Earth's surface. Other ultimate mechanisms can have the same effect (e.g., a large asteroid or comet impact). Both volcanic eruption and asteroid/comet impact are unique, natural processes, but both can cause species extinctions at locations far removed from the site of the volcano or site of the impact by inducing operation of the proximate extinction mechanism, global cooling. Of course, the term 'ultimate' should be regarded as relative when used in discussions of extinction causality. Volcanic eruptions and asteroid/comet impacts are themselves the products of other even larger scale planetary and astronomical processes.

OPPOSITE Eyjafjallajökull fissure-style volcanic eruption, Iceland, April 2010.

PROXIMATE EXTINCTION MECHANISMS

GLOBAL COOLING

Global cooling can be induced by variation in solar radiation (see p.60), by atmospheric composition (e.g. reduction in the proportion of greenhouse gases) or by the proportion of the Earth's atmosphere occupied by clouds – its cloud cover. Clouds reflect solar radiation back into space. An increase in cloud cover results in an increase in the Earth's capacity to reflect incoming solar radiation – its albedo. If increased cloud cover reflects a greater proportion of solar radiation back into space the Earth's average temperature will decline. By the same token as growth in the areal extent of snow and ice fields on the Earth's surface increases, this will also increase the planet's albedo and can cause global cooling. Note that, in this case, the potential exists for the lowered global temperatures to result in greater snowfall in high latitudes, which could result in the establishment of a positive feedback mechanism that promotes runaway global cooling. Finally, sea-level change can lead to global cooling as the dark ocean water absorbs solar radiation. Periods of Earth history during which sea level has stood low tend to be cool times for the planet with strong latitudinal temperature gradients as the exposed continents have a higher reflectivity than the darker ocean waters. Intervals in which sea level has stood high tend to be characterized by warmer average global temperatures with less pronounced latitudinal temperature gradients.

These mechanisms rarely operate in isolation. Increased cloud cover resulting from (say) a volcanic eruption could, in principle, promote the growth of snow and ice fields, which could, if enough water became locked up as snow and ice on the continents, lower sea level, which could increase the planetary albedo, which could result in further global cooling.

SEA-LEVEL CHANGE

Although minor amounts of water are added to the Earth's surface over time through volcanic eruptions and taken away as a result of subduction of sea floor in the deep-ocean trenches, the total volume of water on Earth is relatively constant and has been since the planet's final differentiation into core, mantle and crustal zones was completed and the local ice-rich comets and asteroids caught in the Earth's gravitational field were swept up early in the planet's history. Nevertheless, the distribution of rocks containing marine fossils or traces thereof on the continents makes it clear that, during many intervals in Earth history, sea level stood literally hundreds of metres higher than it stands today (see p.52). Indeed, the modern world is characterized by a lower sea level than at almost any other time in Earth history. In some cases these fluctuations are the result of local or regional changes in the level of land such as those caused by the uplift of large mountain ranges or the rebound of the land surface after the melting of a large ice-cap. But these mechanisms cannot account for changes of more than a few tens of metres or the fact that many large sea-level rises and falls appear to have occurred synchronously on all continents.

Two mechanisms for achieving large, synchronous changes to sea level have been proposed to date. The most obvious, most common, of these is continental glaciation and/or glacial melting. If the average precipitation in high-latitude or high-altitude regions exceeds the average melting of snow and ice these materials will build up over time. Accumulations of snow compact over time to form ice, which is only semi-solid and can flow like a very thick, viscous syrup when put under the pressure of large piles of snow/ice accumulation. Large bodies of glacial ice formed in this way can also, over time, flow over the land under the influence of gravity. In extreme cases glaciers can cover substantial portions of continents as they do today in Antarctica, Greenland and Scandinavia as well as parts of both Canada and Russia. The water that comprises these glaciers ultimately comes from the oceans. During times of widespread continental glaciation the level of the oceans can fall hundreds of metres. Similarly when warm environmental conditions return and the glaciers melt, the level of the oceans can rise by hundreds of metres with the rising ocean waters submerging broad areas of the continents. The timescales over which this process plays out vary, but can easily involve tens of thousands to as much as a million years.

Glaciation is usually the preferred explanation for the common geological observation of sea-level fluctuations that take place over geologically short time intervals. This mechanism works if the magnitude of sea-level change is low. Nevertheless, it is difficult to see how large changes in sea level (e.g. hundreds of metres or feet, which imply the existence of polar ice caps and/or continent-wide glacier fields) can be explained using this model unless there is independent evidence of widespread glaciation (e.g. glacial striations, glacial geomorphological features, dropstones, cold-proxy sediment accumulation patterns, isotopic evidence). Still, in times remote from our own, that are characterized by limited outcrop areas, the absence of primary isotopic signatures and uncertain stratigraphic correlations the absence of independent forms of evidence is perhaps understandable. Absence of evidence is indeed not evidence of absence. Moreover, the documentation of rapid sea-level change by sedimentological and/or palaeontological data is itself a form of evidence.

The other generally accepted mechanism that can result in significant changes in the level of the sea involves changes to the volume of the ocean basins. Local changes to the volume of small basins can arise as a result of local processes (e.g. earthquakes, landslides). But large changes of this type are thought only to be able to be produced by processes related to plate tectonics, either changes in the rate of heat flow at mid-ocean ridges or regional deformations of the Earth's crust that occur at ocean plate subduction centres. If heat flow increases at the mid-ocean ridges the ridges swell causing the volume of the ocean basins to diminish. Reduction in the amount of space available in the world's ocean basins can cause sea level to rise, flooding the coastal areas of the continents. In some cases this process can lead to the formation of large, shallow seas occupying the interiors of continental plates, so called epeiric or epicontinental seas. If heat flow decreases,

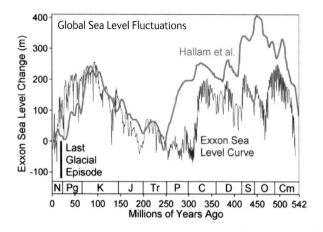

ABOVE Illustration of the Exxon (Haq 1991) and Hallam (1992) sea-level reconstructions for the Phanerozoic (blue and red lines respectively).

the mid-ocean ridges contract, the volumes of the ocean basins increases and sea level falls draining the continental interiors and oftentimes exposing the continental shelves. At the margins of an ocean basin, if subduction of an ocean plate is initiated by the formation of a deep-ocean trench the volume of the ocean basin will increase, causing sea level to fall. If subduction ceases the deep-sea floor in the vicinity of the subduction centre rises causing the volume of the ocean basin to decrease and sea level to rise. The magnitude of sea-level changes resulting from these mechanisms is thought to be less than that from glaciation, but the timescale over which these tectonically induced changes occur is similar to that of continental glaciation.

As can be seen from the reconstruction of Phanerozoic sea-level patterns in the graph above, sea level has been highly variable and organized into both short-term and long-term trends. Hallam (1992) and the Exxon group (see Haq 1991) used very different techniques to assess global sea-level changes. Hallam 's approach is qualitative and uses regional-scale observations from exposed geological sections to estimate the areas of flooded continental interiors. Exxon's approach relies on the interpretation of seismic profiles to determine the extent of coastal or marine sediments of different ages deposited on the continental margins. The Hallam (1992) curve is less sensitive to rapid fluctuations in sea level. The Exxon curve tends to over-represent local geological changes resulting in bias towards the generation of seemingly rapid fluctuations. The depth scale in the image above is as reported by Exxon. Both curves are adjusted to the 2004 International Commission on Stratigraphy (ICS) geological timescale. Because Hallam's sea-level change pattern is reported as an uncalibrated curve, the sea-level change scale shown above pertains only to the Exxon curve.

Aside from their role in inducing changes in average global temperatures and initiating regional or global anoxia events, sea-level changes can cause extinctions as a result of reducing the areas of either shallow marine or terrestrial habitats available for colonization by species. Under these conditions mobile species are forced to migrate with their preferred habitats. If these habitats disappear local populations can become extinct. Populations that are able to track their shifting habitats can also become overcrowded as habitat areas diminish, and species may be brought into contact with novel competitor species, new diseases, etc., all of which can alter their extinction probability. Species that cannot migrate (e.g. reef framework species, many plants) must either adapt to the new conditions or attempt to track their shifting habitat using reproduction-linked dispersal mechanisms. The relation between species-richness values and the amount of habitable area can be modelled quantitatively. This is known as the species-area effect (see opposite).

LEFT The Broken Island region of British Columbia. Islands of different sizes are able to support different numbers of species.

BELOW Species-area effect as illustrated by a graphic analysis of species census data from a series of contiguous habitats. Plot of sample areas vs number of species on an arithmetic axes illustrating the non-linear character of the typical species-area pattern (left), and plot of the same data after logarithmic transformation illustrating conformance to a generalized power function: $S = cA^z$ where A is the area, S, the number of species, and both c and z are constants.

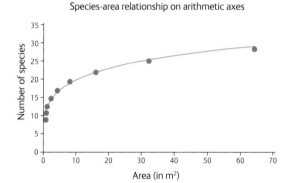

Species-area relationship on arithmetic axes

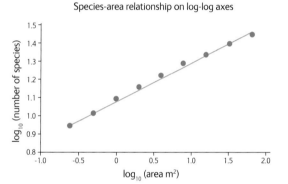

Species-area relationship on log-log axes

In addition to these effects, sea-level changes also result in changes in the composition of atmosphere as a result of the exposure or burial of reactive chemicals usually created as by-products of the organic decay processes. The exposure of continental shelves during a sea-level lowstand tends to release greenhouse gases into the atmosphere, thereby inducing global warming. The submergence of continental shelves during a sea-level highstand tends to sequester greenhouse gases in the sediments and so induces global cooling.

Finally, changes in sea level promote the amelioration or exacerbation of the occurrence of extreme environmental fluctuations on the continents, a phenomenon known as continentality. Because of the high heat-storage capacity of water, ocean temperatures are much more stable than the temperatures of the rocks that make up the continents. Islands and regions of continents located close to coastal areas experience less intense summer-to-winter temperature variations than are present in the interiors of large continents, which also tend to receive less precipitation than coastal zones. If sea level falls markedly the areas of continents increases and their interiors become more isolated, thus increasing their dryness and annual temperature fluctuations. Sea-level highstands result in flooding of the continents – often quite extensive flooding – which serves to reduce the continentality effect. The extent of the continentality effect is also influenced by the intrinsic size and distribution of the continental landmasses, which is ultimately under plate tectonic control.

Continentality changes do not only affect organismal populations directly. They can promote or diminish the creation of high-pressure zones in the Earth's atmosphere which, in turn, can change the track of the atmospheric jet streams and alter the character of atmospheric circulation. Similarly, changes in sea level can also change the character of marine circulation and so alter the processes that distribute heat over the entire planet (see below).

MARINE ANOXIA

In order to survive, marine animals must have access to oxygen that is dissolved in seawater. The only source of this oxygen is diffusion into the marine water column from the atmosphere across the ocean's surface. Because of its proximity to the source of oxygen, marine surface water is almost always well oxygenated. However, transportation of oxygenated waters to the deeper parts of the marine water column depends on vertical marine circulation patterns. Anything that disrupts this circulation or increases the rate of oxygen usage at depth can potentially create zones of low oxygen (disaerobic) or no oxygen (anaerobic or anoxic) conditions in deep-marine waters.

In most ocean basins less dense, well oxygenated, surface waters are separated from denser, less oxygenated, deeper waters by a transition region called the pycnocline. A strong density contrast between these two zones can effectively serve as a cap, inhibiting vertical circulation and promoting the depletion of dissolved oxygen at depth. The pycnocline can be strengthened by temperature and/or salinity differences between surface and deeper marine waters. In addition to this physical effect, the transportation of oxygen to deeper waters can be affected by biological factors, essentially the number of organisms using dissolved oxygen both in the surface waters and at depth. If surface productivity is high most of the available oxygen in the surface waters will be taken by the animals living there. If deep water biomass is also high there won't be sufficient oxygen to support it, especially if vertical circulation is sluggish or reduced. This mechanism can be strengthened if the amount of nutrient materials being shed from the continents into the ocean basin increases. As the rate of nutrient delivery to the ocean basin increases as rates of chemical erosion on the continents increase, that formation of anoxic deep marine waters tends to increase with global warming and decrease with global cooling. These mechanisms also influence rates of marine circulation by decreasing or increasing temperature contrasts between low latitude and high latitude region (see p.58).

The presence of low-oxygen conditions is usually signalled in the rock record by the deposition of organic-rich, dark brown of black, fine grained sediments which, over time, are turned into the rock shale. In well-oxygenated environments dead organic matter is oxidized by the dissolved oxygen in the water – often helped along by the actions of bacteria and tiny scavengers. But under low or no oxygen conditions this organic matter is prevented from decaying and so simply builds up over time.

Organisms endemic to deep-ocean environmental often exhibit adaptations that allow them to tolerate low-oxygen conditions. However, if for some reason this zone of low-oxygen or anoxic water shifts its position suddenly – for instance rising up out of the deep ocean basins as a result of sea-level rise (see p.51) and displacing the previously well-oxygenated waters of the continental shelves, the radically changed environmental conditions can lead to the wholesale extinction of the shelf biota which is composed of species adapted to living in waters with abundant dissolved oxygen (see Chapter 15 for additional discussion).

OCEAN–ATMOSPHERE CIRCULATION

The final extinction mechanisms we will consider are ocean and atmospheric circulation patterns. Both systems are affected by, and to a large extent control, the manner in which heat is distributed over the Earth's surface. These systems are largely stable in terms of their gross organization, but can be affected by a variety of geological factors, several of which fall into the category of ultimate extinction causes (e.g. plate tectonics, large igneous province volcanism and bolide impact, see below). Moreover, ocean–atmosphere circulation patterns are themselves responsible for the climate and the weather found at any given location on the Earth's surface. These are primary determinants of where plant and animal species live on that surface as well as the geographic history of plant and animal species' occurrence patterns over time.

Atmospheric circulation is the more active system. In both hemispheres wind belts are organized into three parallel bands: the Hadley Cell, the Ferrel Cell and the Polar Cell (see p.56 top). In the Hadley Cell warm, moist air ascends near the equator because of its lower density. As this air rises it expands and cools, shedding its moisture as rain. In the northern hemisphere this air flows north as a series of northwesterly high-altitude winds, and draws cooler, drier air into the tropics behind it as a result of the low-pressure zone created by the rising air mass. This cooler, drier air descends around 30˚N latitude where it reverses direction and becomes southeastward flowing middle-altitude, cool, dry air (the Trade Winds), thus completing the cell. The Hadley Cell circulation system in the southern hemisphere is the mirror image of that described above.

The Polar Cell operates in an identical manner to the Hadley Cell. Here relatively warm, moist air ascends into the lower zone of the atmosphere (the troposphere) at around 60˚N and 60˚S latitudes, drawing warmer, middle-latitude air in behind by virtue of the low pressure created by the ascending air mass. This circulation pattern establishes a set of middle-altitude easterly winds. The rising air cools via expansion and sheds its moisture as either rain or snow. The cooler dryer air flows polewards as a series of westerly high-altitude winds where it descends to form a northeasterly or southeasterly flowing middle-altitude air mass, thus completing the cell. Again, the Polar Cell circulation system in the southern hemisphere is the mirror image of that described above.

While the Hadley and Polar cells are both well-defined, mostly closed circulation systems, the Ferrel Cell is an open system that serves as a thermal and pressure buffer between them. The Ferrel Cell overrides the Hadley and the Polar cells in

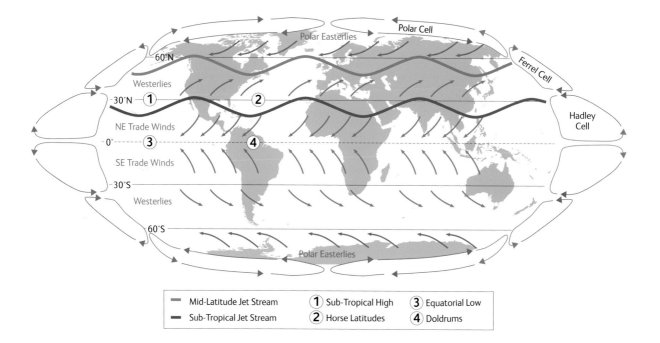

	Mid-Latitude Jet Stream	① Sub-Tropical High	③ Equatorial Low
	Sub-Tropical Jet Stream	② Horse Latitudes	④ Doldrums

ABOVE Atmospheric circulation patterns. Note the idealized three-cell atmospheric circulation pattern generated by a rotating Earth-like planet. The clockwise deflections of middle and high-altitude winds in the northern hemisphere, and anti-clockwise deflections in the southern hemisphere, are caused by the Coriolis Effect. Also shown are the subtropical and mid-latitude jet streams in the northern hemisphere. A similar set of jet streams are present in the southern hemisphere. See text for additional explanation.

RIGHT Results of a computer simulation showing annual mean surface temperatures in degrees Celsius at the time of the Permian extinction.

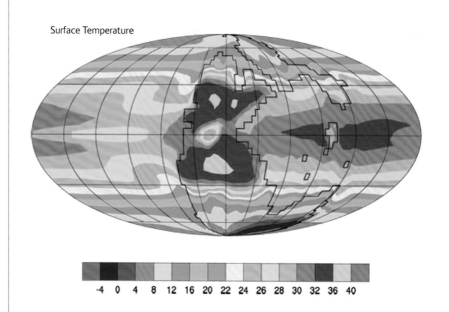

Surface Temperature

both hemispheres and is sometimes referred to as a zone of mixing. Ferrel Cell wind directions can be variable depending on local Hadley and Polar cell intensities, but tends to have a prevailing westerly character. This is the zone in which the jet streams operate. These are high-altitude, narrow, high-speed air currents that occur in the boundary between warmer and cooler air masses at the Hadley/Ferrel cell boundaries, around 30˚N and 30˚S latitude (the subtropical jet), and between the Ferrel and Polar cell boundaries around 60˚N and 60˚S latitude (the polar jet). As the contrast between warm and cold air masses is greater in the northern latitudes the polar jets usually exhibit higher wind speeds than the subtropical jets.

The three-cell structure of the Earth's present atmospheric circulation system is presumed to have been a feature of the Earth since its atmosphere was first established. During warm phases of Earth history simulation studies indicate the cells shift polewards and there tends to be greater intensity difference between the Hadley and Polar cells along with their associated jet streams. During cool phases the cells shift towards the equator and tend to be more equable. During times of extremely high continentality (e.g. Late Permian) a fourth arid cell may have appeared in the equatorial regions driven by convection (see opposite), but this is by no means clear. Even if this arid cell has been part of the Earth's atmospheric organization from time to time, simulations indicate that the basic Hadley–Ferrel–Polar cell structure remained intact.

Atmospheric circulation patterns have a strong influence over the manner in which heat and precipitation are distributed over the Earth's surface. Moreover, these patterns can be changed by a variety of external influences. The tectonic uplift of large mountain ranges, the existence of extensive ice-fields and large volcanic eruptions are also known or suspected to affect the details of atmospheric circulation patterns, especially the location of the jet streams.

Like the Earth's atmosphere, the Earth's oceans also exhibit a three-dimensional circulation pattern. Circulation of waters near the ocean surface is driven by a combination of wind drag and physical forces and shaped by the distribution of continental landmasses across the planet's surface (see p.58 top). Generally, surface currents in the low latitudes are driven westward by the subtropical easterly winds that arise as a result of Hadley Cell circulation. When these currents encounter a continental landmass (e.g. Central and North America in the northern hemisphere, South America in the southern hemisphere) they turn north or south respectively under the pressure of the wind-driven water behind them. This gives rise to narrow, fast-moving western boundary currents of which the Gulf Stream is an example. As these western boundary currents move north the Coriolis Effect — the apparent deflection of moving objects when they are viewed in a rotating reference frame — drives them eastward, across the ocean basins at middle latitudes where their motion is reinforced by the prevailing westerly winds of the Ferrel Cell. Upon encountering a landmass on the eastern side of the ocean basin, part of the current turns south (e.g. along the western coast of Europe and Africa), where it rejoins the easterly flow of surface waters driven by the Trade Winds. If a channel exists to the north, part of the current flows in that direction, where it joins the polar current systems which might be complex and influenced by the presence of local landmasses (e.g. in the present-day northern hemisphere) or adopts a simple circumpolar current system if no landmasses exists to impede its flow (e.g. present-day southern hemisphere). The balance of water across the system is maintained by a series of countercurrents (e.g. North and South Equatorial countercurrent, South Subpolar Countercurrent). Thus, broadly circular surface current systems, or gyres, are maintained in each ocean basin. These gyres circulate clockwise in the northern hemisphere and anticlockwise in the southern hemisphere under the influence of atmospheric circulation patterns and the Coriolis Effect.

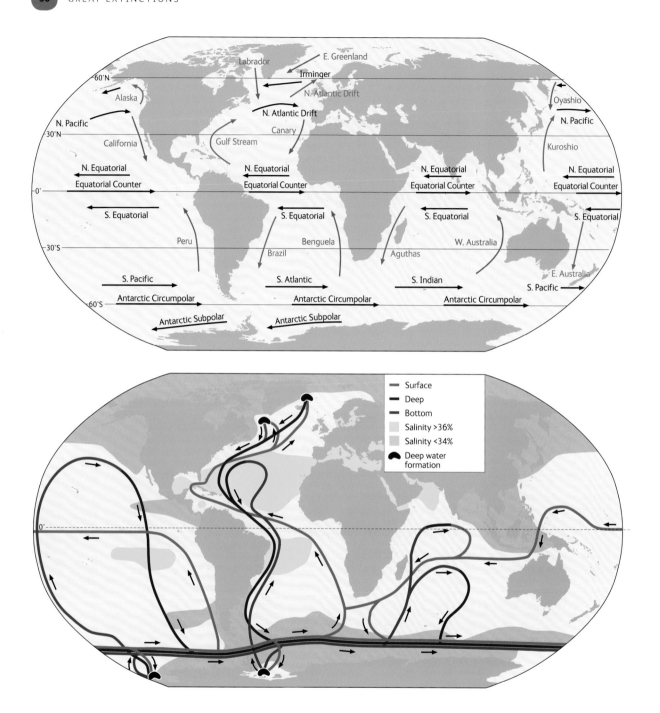

Vertical circulation in the modern oceans is driven by a combination of temperature and density contrasts (see above). The warm, relatively fresh water of the Gulf Stream along the eastern coast of North America cools and grows denser due to evaporation on its way across the Atlantic Ocean. In the area around Greenland and Iceland part of this water sinks and becomes North Atlantic Deep Water (NADW), passing through a series of submarine canyons in this region and finally emerging as deep water in the Atlantic Ocean Basin proper. From this point it flows south down the length of the Atlantic Ocean Basin to a point where it is forced back to the surface under pressure from even colder and denser Antarctic surface waters that are sinking in this region.

Bottom-water flow of this sort occurs in all modern ocean basins, but is under the control of deep-ocean passages. Prior to the existence of these deep-ocean passages – which became established in the modern oceans 10–15 million years ago it is presumed there was no bottom-water circulation. In the absence of such circulation patterns deep-ocean waters would be expected to grow stagnant or anoxic as the dissolved oxygen they contain is depleted through the oxidation of organic materials. During such intervals marine species not specifically adapted to living in low-oxygen environments can be driven to extinction. Even more importantly, a deep-ocean basin full of anoxic water during a sea-level lowstand can provide a source of corrosive, anoxic bottom water that can rise to cover shallow marine areas – causing widespread extinctions of marine bottom-dwelling species – if sea level rises rapidly.

OPPOSITE Oceanic circulation patterns. Top, marine surface currents. Red arrows indicate relatively warm currents, blue arrows indicate relatively cold currents. Below, vertical marine currents. Red bands indicate relatively warm, shallow currents. Blue bands indicate relatively cold, deep currents. Purple bands indicate very cold bottom water flow currents. Yellow icons mark positions of deep-water formation as cold, polar surface waters sink to become marine bottom water.

ULTIMATE EXTINCTION MECHANISMS

SOLAR RADIATION

Although Earth's sun is often regarded as a metaphor for constancy, the output of solar radiation has been surprisingly variable throughout recorded history and, as assessed by proxy observations, beyond (see below). Historical periods of lower-than-normal temperatures in 1460–1500 (the Spörer Minimum), 1645–1715 (the Maunder Minimum) and 1790–1830 (the Dalton Minimum) are all thought to have been caused by variations in solar radiation. Although it is not possible to measure the rate of solar radiation directly for geologically ancient times, it is reasonable to suspect that similar variation in the sun's energy output characterized the geological past. Over the longer term theoretical models suggest that during the last 2.5 billion years the sun's energy output has increased by as much as 25%.

BELOW Patterns of variation in two solar activity proxies as recorded over historical time. Left, cosmogenic beryllium isotopic composition (blue) and sunspot number (red). Right, inferred pattern of variation in solar radiation over the last 10,000 years.

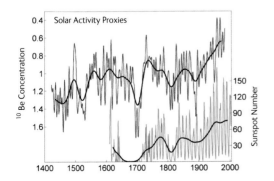

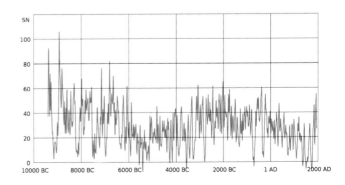

PLATE TECTONICS

There are many ways plate tectonic processes can cause, or contribute to, extinctions. When continents collide with one another they increase their area and isolate their interiors from the ameliorating affects of a nearby ocean basin. This 'continentality effect' makes climates in the continental interiors more extreme: colder in the winters, hotter in the summers. Continental collisions also throw up high mountain ranges that can cause changes in atmospheric circulation patterns – especially the paths of jet streams – starve the continental interiors of rain turning them into deserts and even effect marine circulation intensities. If a large continental landmass moves over a polar region, the continentality effect can set up a cold-trending climate feedback loop that may result in glaciers spreading across the surfaces of continents, resulting in global cooling and a lowering of sea level. Similarly, if a continental landmass previously located over a polar region moves away from the pole as a result of tectonic activity, continent-sized glaciers can melt causing average global temperature and sea level to rise, flooding the continents and diminishing the land surface area available for colonization by terrestrial species while increasing the oceanic area available. Tectonically driven sea-level changes and glacial advances/retreats also have implications for the planet's albedo and the composition of its atmosphere as greenhouse gases can be released from, or sequestered in, natural stores (see p.53). Finally, the heat flow associated with plate tectonic processes can change the volumes of ocean basins further affecting sea level, while increased or decreased volcanic activity can release (or reduce) compounds that affect the acidity of rain and snow.

LARGE IGNEOUS PROVINCE (LIP) VOLCANISM

Large Igneous Province volcanism is a special class of volcanic eruptions usually associated with a stable point or tectonic 'hotspot' on the Earth's surface. Such eruptions result in large outpourings of lava, typically of a basaltic character (see opposite). Because of their geographically stable nature these hotspots are felt to be only loosely connected with standard plate tectonic processes though their origin remains something of a mystery. Long-lived hotspots (e.g. the Hawaiian hotspot, the Réunion hotspot, the Yellowstone hotspot) can leave a trail of volcanic accumulations on oceanic and continental tectonic plates as these move over the hotspot during the course of millions of years. These eruption events can emplace millions of cubic kilometres of particulate lava and ash and release enormous quantities of associated corrosive gases, carbon dioxide (CO_2) and water into the atmosphere within relatively short geological timeframes (e.g. 50,000–1,000,000 years). Eruptions of this magnitude would be catastrophic for the local environment and perturb a variety of planetary factors (e.g. albedo, rain acidity, chemical composition of the atmosphere, presence of greenhouse gases, atmospheric circulation patterns), all of which can have an effect on organismal populations worldwide (the SC scenario). Moreover, the possibility exists that, if LIP volcanism occurs at a time when species are already under environmental stress (e.g. during a sea-level lowstand with the ocean basins in an anoxic condition), the effects of two unrelated environmental processes can combine to produce a greater-than-normal level of extinction intensity (the MIC scenario).

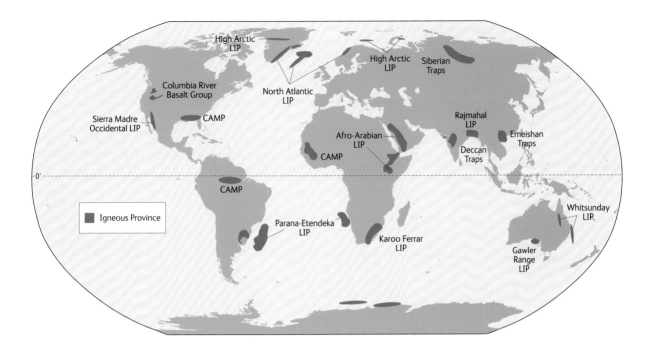

ABOVE Large igneous province (LIP) volcanic eruptions and regions in which LIP eruptions have taken place at various points in Earth's history.

LARGE BOLIDE IMPACT

For quite a long time it was assumed that collisions between the Earth and large extraterrestrial objects were something that happened in the Earth's early history, but that the solar system had been swept clean of all such objects billions of years ago. This assumption was proven false when Eugene Shoemaker demonstrated that the circular structure now known as the Barringer Meteor Crater in Arizona resulted from a collision between the Earth and a meteor as little as 50,000 years ago (see p.14). Later, a team led by the Nobel Laureate Luis Alvarez demonstrated that a large bolide (a generic term for a large extraterrestrial meteor, comet or asteroid) had collided with the Earth close to or at the time of the end-Cretaceous extinction event, some 65 million years ago. The initial evidence for this impact event was a concentration of the rare earth element iridium in the sediments of a Cretaceous–Palaeogene (K–Pg) boundary section in Italy. Subsequent research, uncovered other types of evidence for the K–Pg bolide impact, including shocked quartz crystals and impact glass, at several widely distributed K–Pg boundary sections and cores (see Chapter 12). Once the principle that impacts have occurred at various times in Earth history was established, enigmatic circular physiographic features across the world were reappraised to determine whether they too could be impact craters. To date almost 200 known or suspected impact craters have been discovered ranging from 0.135 to 300 km (0.884 to 186 mi in diameter) (see p.62). No doubt many more impacts have occurred in Earth history, but their craters have eroded away or have been buried by subsequent sediments. Accordingly, the incidence of bolide impact can no longer be considered an unusual event geologically speaking.

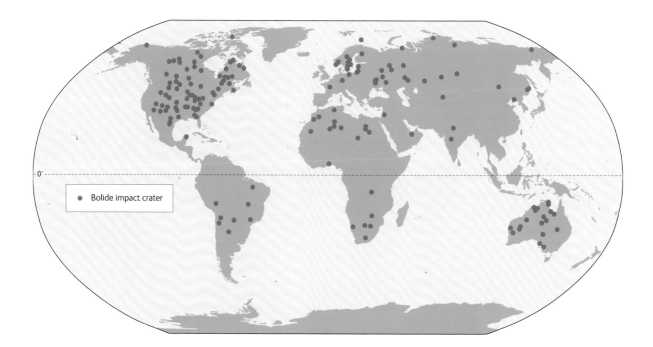

Bolide impact crater

ABOVE Map of 182 confirmed bolide impact structures as listed by the Earth Impact Database, Planetary and Space Science Centre (PASSC), University of New Brunswick.

RIGHT Landsat photograph of the Manicouagan crater, Côte-Nord region, Québec, Canada.

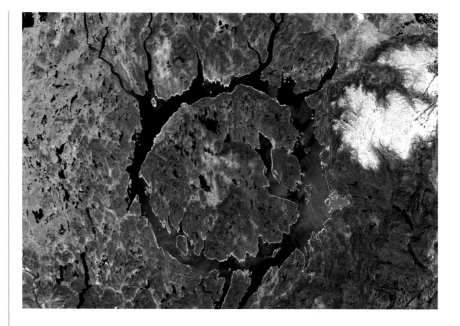

Large impacts such as the Chicxulub event, which is regarded by most (but not all) researchers to have occurred coincident with the K–Pg boundary, would have a variety of very intense but relatively short-lived effects (the SC scenario). These include a thermal flash, which would have incinerated all exposed flammable material for a considerable distance around the impact site possibly leading to the ignition of wildfires hundreds (or even thousands) of kilometres away; a shock wave, which would have destroyed large structures; exceedingly large earthquakes and tsunamis (the latter of which would have devastated many coastal areas); global darkness as a result of material thrown from the crater high into the Earth's atmosphere; increased planetary albedo with associated short-term global cooling; longer-term global warming due to release of large quantities of greenhouse gases; and increased acidity of rainwater due to chemically reactive materials ejected into the atmosphere from the crater. Simulations suggest the global effects would, at most, have lasted 100 years. Material in the crater itself would have remained hot for a much longer period with consequent potential for perturbation of local or regional heat flow and/or weather patterns. As with LIP volcanism, if a large bolide impact event occurred at a time when the biosphere was already under environmental stress (e.g. during a sea-level lowstand with the ocean basins in an anoxic condition, an LIP volcanic event, global warming) the effects of unrelated environmental processes could combine to produce a greater-than-normal level of extinction intensity (the MIC scenario).

6 Precambrian and Cambrian extinctions

THE PRECAMBRIAN EXTINCTION

OPPOSITE Cambrian sandstones exposed in Rocky Arbor State Park, Wisconsin Dells, USA.

The earliest unquestionable fossils date from 3,500 million (3.5 billion) years ago. These have the form of thin organic mats called stromatolites, similar in character to mats of blue–green bacteria found along some marine coastlines today. Modern stromatolite mats secrete a sticky mucous that traps small rock and mineral particles on their surface. As the layer of sediments accumulating on top of the mat grows thicker, the bacterial colony grows up through the spaces between the mineral grains, thus incorporating layers of hard sediments into the colony's structure. It is this layering of sediments that gives the stromatolite its distinctive form and is responsible for preserving the colony during fossilization.

Prior to this time it is assumed life existed in even more primitive states. In the modern world various types of quasi-lifeforms exist (e.g. viruses, virusoids, viroids, prions) that are able to engineer their own reproduction by a variety of mechanisms. Some primitive lifeforms are thought to have become incorporated as the organelles of later cells (e.g. ribosomes, chloroplasts) and so to persist literally as living fossils. But leaving these matters aside, the oldest undoubted organisms preserved in the fossil record exhibit a fairly advanced, bacterial level of structural organization.

ABOVE AND LEFT Above, fossil stromatolite from Bolivia. Left, stromatolite-like structures found at Shark Bay, Western Australia. Microscopic examination of these structures shows that they are composed of numerous cyanobacteria (formerly blue-green algae) mats that have been preserved as a result of their ability to trap sediment and form large stony cushion-like masses. These masses are among the oldest organic remains to have been found, ranging from 2,000 to 3,000 million years old. In coastal waters such as Shark Bay stromatolite formation continues today.

Life at the bacterial grade of complexity existed for some 2,500 million years, adapting via natural selection to different environments and changing environmental conditions during what is, in fact, the greater portion of the Earth's existence. Then, towards the end of this Proterozoic (= primitive life) Eon an experiment in constructing a more complicated organism succeeded to the extent that examples of, to our eyes, much more complex, yet strange, creatures enter the fossil record. This group of unusual, multicellular creatures is known as the Vendian or Ediacaran biota. In a series of interesting articles the German palaeontologist Adolf ('Dolf') Seilacher has shown that these sessile, benthic, marine organisms were characterized by entirely soft bodies and are preserved only as impressions in the sediments that surrounded them when they died (see below). Based on these impressions Ediacaran organisms appear to have been constructed out of tube-like compartments and, overall, exhibited a frond-like or tube-like form, often with a quilted surface texture.

Little is known of evolutionary relations among members of the Ediacaran biota or how they fit into the overall history of life. Some Ediacaran fossils suggest sponge-like or coral-like affinities. Others superficially resemble primitive molluscs. These were the most complex and arguably successful species on the planet for some 70 million years, until they disappeared – perhaps relatively quickly – near the top of the Proterozoic in what is probably the oldest mass-extinction event.

Because of the organisms' extreme age and soft-bodied character, details of the Ediacaran biota's extinction are very sketchy. What we do know is that no unquestioned fossil that exhibits the characteristic Ediacaran body plan has ever been recovered from sediments younger than 540 million years old and that

RIGHT A *Dickinsonia* fossil from Australia. *Dickinsonia* is a commonly found member of the Ediacaran biota.

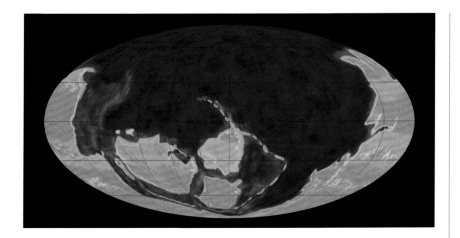

LEFT Early Cambrian palaeogeography (*c.* 540 million years ago).

no Ediacaran-like animals exist today. Nonetheless, at the base of the Cambrian Period a large number of more advanced multicellular body types appear, body types that persist for hundreds of millions of years in the Phanerozoic (= visible life) Eon and some of which remain within the modern world. Many researchers believe the apparent rapidity of the Ediacaran extinction may be misleading and controlled more by preservational factors than biology. Nevertheless, it is difficult to avoid the impression that the appearance of these new and radically different organisms, and the disappearance of the Ediacaran biota, are linked causally. It is also known that at the end of the Proterozoic the Earth was experiencing many profound changes in its physical environment, including the break-up of the Proterozoic supercontinent Rodinia (see above) and rising sea levels worldwide, along with fluctuations in both ocean and atmosphere composition and circulation patterns.

Because we remain a long way from understanding the timing and the causes of the Ediacaran extinction, we are currently unable to discount any of these factors as playing a role in either causing, contributing to, or mediating this biotic transition. To my mind this makes an important point that is often overlooked in the debates over younger, better known extinction events - when faced with the absence of any data that can identify a specific cause or set of causes for an extinction event the best course to action is to remain open minded regarding the cause(s).

In the geological literature this approach is enshrined as the method multiple working hypothesis approach first described formally by T. C. Chamberlin (1897). Both the SC and MIC scenarios are more specific statements of causal relation than the one advocated by Chamberlin. In this sense then, acceptance of either requires positive evidence to be collected and presented in its favour. Because of its remoteness in time this process of causal hypotheses evaluation is at its very earliest stage in the case of the Ediacaran extinction. But here we see an example of the state our understanding of the cause(s) all of the great extinction events were in at one time.

THE CAMBRIAN BIOMERE EXTINCTIONS

The Cambrian is the oldest time interval of the Phanerozoic, which is where the story of life as recorded in the fossil record first starts to get really interesting. This interval is distinguished from older rocks solely on the basis of the fossils that occur in the sedimentary rock layers deposited at this time. It is here that most of the major marine animal phyla that will go on to dominate the seas, the land and the air make their first appearance in the fossil record, notably molluscs, brachiopods and arthropods. Other groups of marine invertebrates that are less familiar – because they are extinct – also make their first appearance near the base of the Cambrian, including halkieriids and hyoliths.

SETTING

Because the first representatives of each of these groups already exhibit a remarkably advanced state of structural complexity, it is assumed that each had a long pre-Cambrian evolutionary history about which palaeontologists know little. However, these advanced, highly organized, multicellular animals seem to appear in the fossil record out of nowhere because, for some reason, each developed the ability to secrete a hard, external skeleton that was fossilizable at about the same time in Earth history. The apparently sudden appearance of so many different organismal body plans in a relatively short span of time has come to be known in recent years as the 'Cambrian explosion' in biodiversity.

Among the most prominent constituents of the Cambrian evolutionary fauna were the trilobites. These shallow marine, benthic arthropods were literally the cockroaches of their day, occurring commonly in a wide variety of environments and displaying an amazing diversity of shapes and sizes. Because of their common occurrence in sediments of Cambrian age, trilobites have received intensive study as classic Cambrian index fossils. Biozones defined on the basis of trilobite species are used worldwide to infer time relations and correlate packages of Cambrian sedimentary rocks.

EXTINCTIONS

The first notable extinction of the Cambrian occurred at the transition between the Early and Middle Cambrian Epochs. In this interval a major group of primitive trilobites (the ollenellids) and an important group of ancient reef-building organisms (the archaeocyathids) vanish from the fossil record (see opposite). Both groups were major constituents of Early Cambrian shallow marine faunas. Indeed, the appearance of archaeocyathid reefs in the Early Cambrian is an important development in the ecological history of the oceans as these were the first reefs created by animals. [Note: the stromatolite colonies as well as other types of microbial structures found in Precambrian sediments are also, technically, reefs.] The extinction of the archaeocyathids and the olenellid trilobites deserves more study to address the

LEFT AND BELOW Archaeocyathid reefs of the Forteau Formation exposed in southern Labrador, Canada. The metre stick is 1.1 m (3.5 ft) long. Archaeocyathids were victims of the first Cambrian extinction event. A single archaeocyathid individual is shown below.

issue of what caused their demise and how long it took to complete, both of which are unknown at present. Irrespective of its cause, this extinction initialized a pattern of successive decimations of shallow marine animal reef-based habitats that will become an important feature of the extinction record from Early Palaeozoic time up to, and including, our own.

It was while engaged in the biostratigraphic study of Cambrian trilobite faunas in the early 1960s that Allison (Pete) Palmer noticed an interesting pattern in the distribution of Middle and Late Cambrian trilobite species he was studying in the western USA. After plotting the stratigraphic occurrences of trilobites he had collected Palmer noticed that several of the biozone boundaries were coincident with profound changes in the trilobite fauna such that scarcely any of the species present below the boundaries were found in the succeeding biozone. Careful collecting, often on a centimetre-by-centimetre basis, allowed Palmer and his colleagues to identify three such horizons in the Great Basin of the western USA (see p.70). In 1965 Palmer christened these intervals biomeres (= segment of life) and argued that they constituted a new kind of biostratigraphic unit ...

"... bounded by non-evolutionary discontinuities in the fossil record at which taxa of family or lesser rank within one phylum or class were eliminated. These discontinuities [are] separated in time by several millions of years. Above each discontinuity, which was not reflected in the rocks, there was an invasion of elements of a slowly evolving lineage, because successive invaders were very similar to one another. Between the discontinuities, the trilobite assemblages showed only gradual change."

(PALMER 1984, p.599)

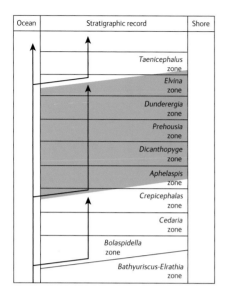

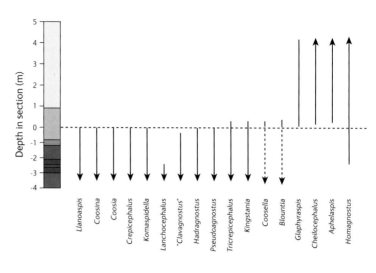

To date as many as seven biomere extinction events have been identified, collectively spanning an interval of some 30 million years, with the last few biomere extinctions occurring in the Early Ordovician. At least one of these events appears to be global in scope. Palmer thought originally that the primary significance of biomeres was as time markers to achieve cross-continent and intercontinental correlation among sedimentary rock layers. But over time the significance of biomeres to the natural history of extinction has also become apparent. Once Jack Sepkoski tabulated biodiversity across all Cambrian stages and compared the stage-level extinction data across the Phanerozoic it was appreciated that those Cambrian stages that contained biomere boundaries were identified as including among the most intense stage-level extinction events (on a percentage basis) of the past 600 million years.

TIMING

Each of the trilobite biomeres have durations of at least several million years and encompass several traditional biozones. They begin with an initial appearance of species from slowly evolving deeper water trilobite lineages. Over time these lineages gradually diversify through speciation into a complete, indigenous, shallow-water fauna. This diversification continues until disaster strikes and the entire fauna disappears over a very small stratigraphic interval typically consisting of not more than a few centimetres of sedimentary rock.

Usually this disappearance is not accompanied by any change in the character of the sediments surrounding the fossils. Based on this observation, the gross character of the environments in which the trilobites lived is thought not to have changed. In most cases the extinction takes place below a subtle, but marked, discontinuity in the sedimentary rock record. This observation suggests that, at the very least,

an interval of non-deposition, and possibly a period of active erosion of the sea-bottom sediments, is associated with the biomere extinctions. However, in some stratigraphic sections no evidence of a break in sediment deposition has been discovered to date.

While trilobites were the focus of Palmer's original investigation, other organismal groups also suffered extinctions at the biomere boundaries, notably brachiopods and conodonts (above). Subsequent field work by Palmer, James Stitt and others demonstrated that Palmer's biomeres were consistent features of the North American Cambrian fossil record stretching from the eastern mid-continent (Tennessee) to the west (Utah, Nevada) and from the south (Texas, Arizona) to the north (Montana).

CAUSE(S)

Three potential causes have been advanced to account for the North American biomere boundary extinctions. The most well-developed explanation was advanced by Palmer and colleagues and involves a combination of sea-level change and marine anoxia. Certainly intervals in the Late Cambrian and Early Ordovician were characterized by stagnant marine

deep waters. Evidence for this comes in the form of characteristic black shales along the margins of the continental fragments existing from Cambrian times. Several researchers have inferred that Cambrian deep-water stagnation was brought about by the existence of a strong thermocline. There is also evidence for a sea-level drop and subsequent rise associated with some (if not most) biomere boundaries. The cause of this sea-level drop remains something of a mystery; however, there have been recent reports of glacial deposits in Canada and Ireland lower Cambrian age (Landing and MacGabhann 2010). There is also good evidence for a sustained Early-Middle Ordovician sea-level rise associated with rifting of the Iapetus Ocean and so, by implication, to tectonic factors.

To an extent sea-level regression is part of the biomere story; as the seas pulled back towards the ocean basins the amount of habitat available for colonization by benthic marine invertebrates such as trilobites decreased. Species that, in former times, lived stably side-by-side were now forced to migrate to new localities and were, inevitably, propelled into an increasingly intense competition for shrinking pools of resources. Unfortunately, sea-level decline cannot be the only mechanism involved in the biomere extinctions because a pronounced sea-level fall also occurred in the middle of the second biomere. Therefore, the extinctions that took place at the biomere boundary are inferred to have involved other factors as well. Palmer, Stitt and Michael Taylor have suggested a two-stage model in which some (presumably small) number of the biomere extinctions are due to the species-area effect that accompanied the Late Cambrian sea-level fall, but the majority are due to effects that accompany the subsequent sea-level rise.

As was noted in the previous chapter, during a severe sea-level fall that drains the continents the level of the sea shifts to the vicinity of the edge of the continental platforms, where the relatively warm, shallow waters of the continental shelves exist in close proximity to the cold, deep waters that lie beyond the shelves. These remnant populations would have had little defence against the cold, anoxic waters that suddenly engulfed them, as is indicated by the black shale deposits that overly the well-oxygenated and fossiliferous continental shelf sediments in many Cambrian stratigraphic successions. Also, the reduced areal extent of many species means that probabilities of any local population surviving such an extensive and sudden regional shift in environmental conditions would have been small; hence, the widespread and taxonomically wholesale extinctions at the biomere boundaries.

Evidence for this model comes from the subsequent repopulation of the shallow-water habitats in the lower parts of the succeeding biomere. In each case most of the trilobites found in this recovery interval are members of the trilobite family Olenidae. Olenid trilobites appear to have preferentially inhabited colder, deeper waters during the latter stages of the biomere intervals, but occur closer to the continental interiors – presumably in shallower water habitats – during the biomere recovery stage. This group of trilobites also serves as the parent stock from which the subsequent evolutionary radiation of new, shallow-water, warm-adapted trilobite species are derived.

The Palmer–Stitt–Taylor scenario is not a perfect fit to the data at hand. In terms of their favoured mechanisms it is difficult to understand how a layer of warm, well-oxygenated water would not have formed close to the surface of the advancing anoxic water and why more trilobite species could not have tracked this zone of relatively benign conditions as it moved towards the continental interiors. This is especially puzzling in light of the fact that many of these same species survived the previous sea-level fall and would have been free to colonize a vast area of newly submerged habitat that contained no indigenous species with which to compete. In addition, prominent victims of these crises included not only benthic trilobite species but also their planktonic cousins (the agnostid trilobites) whose habitat would presumably have been untouched by the cold anoxic waters welling up from the continental margins. Finally, geologists have no ready mechanism to account for why there would have been repeated pulses of sea-level change through Middle and Late Cambrian time. Glaciation would provide an obvious mechanism, but evidence of continental glaciation independent of that of the sea-level fluctuations themselves has proved elusive. Despite these problems though, the Palmer–Stitt–Taylor multiple-cause scenario remains the best supported of the available models (e.g. sea-level change, global cooling, meteorite/comet impact) that have been proposed as causes of the biomere extinctions. [Note: although sediments at the biomere boundaries have been searched for evidence of extraterrestrial collision (e.g. presence of rare earth element anomalies, shocked quartz) no discoveries have been made that would support the operation of this mechanism at any of these stratigraphic horizons.]

In discussions of mass-extinction events the biomere boundary events are usually discounted because global biodiversity was low at this time and because the biomere pattern is known in detail only from North America at present. Clearly the number of species involved in these events is not as large as the number involved in the large end-Permian or the Mesozoic extinctions. Nevertheless, these Cambrian faunas were, inarguably, the dominant components of life on the planet at this time and biomere-like patterns of Late Cambrian faunal turnover have been reported from other regions (e.g. Southeast Asia, Australia). The successive elimination of well-established Late Cambrian biotas in a series of devastating blows to the Cambrian ecosystem constitute important events in the history of life. In fact, because these extinctions came so early in that history, by eliminating a series of heretofore highly successful marine species groups these events set the stage for, and introduced constraints on, many subsequent evolutionary developments in ways palaeontologists are only now beginning to appreciate.

7 The End-Ordovician extinctions

T HE FINAL BIOMERE EXTINCTION signalled a profound change in the character and taxonomic composition of shallow-marine communities. The trilobite–brachiopod communities that characterize this lowermost Palaeozoic interval never recovered their former numerically dominant position of diversity and abundance during the subsequent Early Ordovician evolutionary radiation. In their place a new evolutionary fauna arose that was to dominate the remainder of the Palaeozoic Era (above). Trilobites and brachiopods were important members of this fauna. But the families that predominated through the Ordovician represented more advanced designs than their Cambrian counterparts. These groups were joined by rugose and tabulate corals (the latter of which, along with tree-like ramose, latticed, fan-like fenestrellid and encrusting bryozoans, formed the cores of extensive reef ecosystems), gastropod and bivalve molluscs, stalked blastoid and crinoid echinoderms and, the top predator of the Early Palaeozoic marine habitats, nautiloid cephalopods with either straight or coiled shells. This fauna exhibited a rapid and virtually unbroken radiation of new species through the first six Ordovician stages that was unprecedented at that time. Then, in the final stage of the Ordovician this supremely diverse and, to all indications, well-adapted fauna was torn apart, suffering what many believe to have been the second most severe of the great extinctions.

ABOVE Reconstruction of a typical Ordovician shallow-water marine habitat. Foreground: the eurypterid arthropod *Megalograptus* disturbing an orthoconic cephalopod (left), an orthicon attacking a group of trilobites (*Homotelus*, centre) and the brittle starfish *Salteraster* (right). Background: the articulate crinoid *Balanocrinus*, solitary and colonial rugose corals, and branching bryozoan colonies. Marine diversity rose to 500 families and reefs developed rapidly especially in the tropical waters that flooded Laurentia.

OPPOSITE An outcrop of Cambrian sediments, Lake Champlain, New York State, USA. The stratigraphic sequence is inverted so the Lower Cambrian dolomites lie on top of stratigraphically younger Middle Ordovician sediments.

SETTING

The taxonomic and ecological character of the Late Ordovician extinction event is complex largely because the post-Cambrian evolutionary radiation produced more complex ecological systems. As with the Cambrian extinctions, a marked contrast existed between those species and species associations that inhabited the broad, shallow seas that occupied the interiors of continental platforms and those that inhabited the deeper and less extensive continental margins. Regional differences in biotic composition – endemism – were much better developed in the former than the latter. Within the fossil assemblages that represent Ordovician shallow marine environments distinct tropical, subtropical and temperate provinces can be recognized. However, these highly diverse and well-structured aggregations of marine invertebrates disappeared over the last few million years of the Ordovician with remnant species being pushed out into marginal, deeper water habitats again. The sediments surrounding these fossils tell a story of a rapid and widespread sea-level decline followed, in most stratigraphic successions, by an erosion surface that marks an interval of terrestrial exposure. Just as was the case with the Cambrian biomere extinctions, the physical and biotic evidence indicates the continental platforms were drained of seawater. But this time the drainage was much more complete as well as being much longer lasting.

EXTINCTIONS

Prominent victims of the Late Ordovician extinction event are shown opposite. Despite their repeated reductions at the Cambrian biomere horizons, trilobites suffered once again in the Ordovician. Theirs is one of the more complex stories of this interval, with members of prominent Cambrian groups being progressively replaced by indigenous Ordovician taxa throughout the course of the Early Palaeozoic interval. Then both groups suffered significant losses during the end-Ordovician extinction event.

Overall some 70% of all trilobite genera extant at the beginning of the extinction interval – and well over 90% of the species – had disappeared by its end. Among the brachiopods, members of the Inarticulata suffered most, undergoing a greater than 50% reduction in genera. Among the articulate brachiopods previously dominant orthid and stropheomenid species were also hit hard. These groups never regained their former numbers during the subsequent Silurian recovery, but rather were replaced, at least in the case of articulate species, by more advanced designs of the pentamerid, rhychonellid, atrypid and athyrid brachiopods.

In the Middle Ordovician seas tabulate corals outnumbered those of the Rugosa. But their short-lived dominance was also destroyed by the Late Ordovician extinction event. From the Silurian to the Permian the dominant coral species, both in terms of species numbers and abundance, were rugose corals.

Like the corals the bryozoans were most diverse in the Ordovician epicontinental seas and so were among the groups most affected by the sea-level fall with a loss of as many as 86% of North American species. In particular, the previously dominant cryptostomatids and treptostomatids-type bryozoans never regained their previous species numbers. Even more serious were the number of echinoderm genera lost, though for most crinoid groups the loss was temporary with species numbers recovering to, or exceeding, their Ordovician values in the overlying Silurian. The exception to this rule, however, were the cystoids, a primitive group of stalked echinoderms that never regained their pre-event richness values.

Species that swam through the water, as opposed to crawling along the bottom, fared no better. Of the eight Ordovician nautiloid cephalopod orders, five of the orders either became extinct outright (one order) or failed to recover their pre-extinction richnesses (four orders). In the case of graptolites, their formerly global distribution pulled back to the tropics. At the height of the extinction total graptoloid diversity was represented by a mere six species. This group did rebound in the Silurian. But like the nautiloids, numbers were reduced to such an extent that there was a substantial danger of the entire group's extinction. A number of other benthic and planktonic microfossil groups also suffered substantial extinctions at this time, including acritarchs, chitinozoans, radiolaria, ostracods and conodonts.

TIMING

All geological and palaeontological indications are that the Early and Middle Ordovician were times of extraordinarily warm climates. This is even more impressive than it might seem for the estimates of the sun's energy output history suggest that, in the Ordovician, about 5% less solar radiation was reaching the Earth compared to the amount reaching it today. This would mean the concentration of greenhouse gases (largely CO_2) in the Earth's atmosphere would have been around 10% greater than they are today just to maintain the planet's ambient temperature. This increased concentration of greenhouse gases could be supplied by either or both of two mechanisms: (1) by increased volcanic activity, (2) decreased draw-down of CO_2 from the atmosphere by phytoplankton with subsequent sequestration in the form of marine carbonate ($CaCO_3$) deposition. Unfortunately the geological record does not provide sufficient information at present to determine which (or either) of these mechanisms is responsible for the inferred high concentrations of greenhouse gases in the Ordovician atmosphere.

This situation changed, however, in the early part of the Hirnantian Age. Physical evidence, from the finding of glacial striations gouged into Hirnantian sediments, to characteristic glacial outwash deposits and glacial dropstones from Africa, the Middle East, Europe, Scandinavia and South America, demonstrate that large ice-sheets were moving over the surfaces of continents throughout this latest Ordovician time interval. This direct evidence for continental glaciation also agrees with the timing

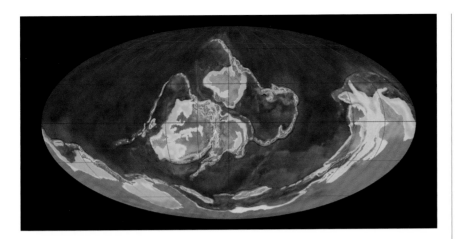

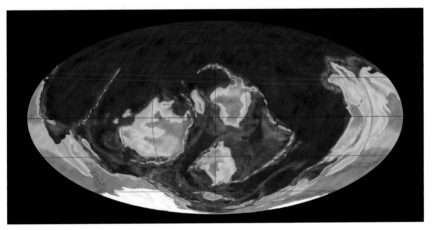

LEFT Lower Palaeozoic palaeogeography. Above, Middle Ordovician (*c*. 470 million years ago), and below, Middle Silurian (*c*. 430 million years ago).

of the Late Ordovician sea-level fall, which may have been as much as 70–100 m (230 –330 ft) in depth.

The extent of glacial cooling was large and its duration protracted, lasting in total for *c*. 1 million years. This Late Ordovician planet-wide cooling was brought about as a result of the distribution of continental landmasses over the Earth's surface. In the Ordovician the present continents of Africa, South America, the Middle East, India and Antarctica were joined together in one supercontinent, Gondwana, which drifted over the South Pole in the Middle and Late Ordovician (see above). The intense glaciation that followed, and equally intense global release from glacial conditions, set up a classic one-two punch to the biosphere that decimated both planktonic/pelagic and benthic organismal groups.

CAUSE(S)

The proximal cause advanced almost universally by geologists and palaeontologists to account for the devastation suffered by so many fossil groups at the close of the Ordovician: intense continental glaciation (Stanley 1987, Sheehan 2001). Outside this consensus, however, there is little agreement with regard to the roles played by various ultimate causal processes. The one constant among the various scenarios that does exist is the observation that, throughout Earth history, whenever there has been a long episode of continental glaciation a continent-sized landmass has been present at the Earth's South Pole. This is not a simple relationship. There have been times when a continental landmass has been located over the pole and the Earth has experienced moderate climates. But having a continent in the southern polar position makes the Earth vulnerable to so-called 'ice ages'. If other factors result in global cooling, a southern polar continent can magnify the effect of those factors and flip the planet into a new and much colder state.

Many geologists who study the Ordovician believe this proximal 'other factor' involved a change in the pattern of deep-ocean circulation. There is good evidence that pre-Hirnantian marine deep-ocean waters were severely depleted in oxygen in the form of the widespread black shale deposits commonly found associated with Ordovician deep-marine faunas. As a mechanism, Ordovician deep-sea anoxia is not especially mysterious. As noted above there is good reason to suspect the Earth's atmosphere contained a greater proportion of greenhouse gases – and so a lower proportion of oxygen – than is the case at present. Under these reduced oxygen conditions anoxic bottom waters would be expected to have developed, especially beneath the areas with high sea surface productivity such as those beneath deep-sea nutrient upwelling zones (e.g. along the eastern margins of continents in the modern oceans). However, with the onset of glacial conditions these black shales disappear. In their place geologists find the brownish and yellowish sediments they associate with sediments deposited in oxygen-rich environments. Even more importantly, in these latest Ordovician deep-marine settings the burrows of marine species are present – traces of species that required oxygen to live.

The diverse Middle Ordovician marine biota was adapted to the environmental conditions that predominated at that time, including the presence of a deep layer of cold, anoxic marine bottom water. There is evidence of whole communities of organisms adapted to low-oxygen conditions occurring in the Middle and Late Ordovician. However, when this state of environmental affairs changed suddenly with the onset of glaciation extinctions would be the expected result. In particular, the removal of the critical bio-limiting nutrient phosphorus from the oxygenated post-glaciation waters is thought to have depressed primary productivity in the Late Ordovician oceans severely. Also, based on analogy with the modern ocean, if Ordovician deep-ocean circulation patterns had changed such that cold, oxygenated waters were being drawn down into the deep sea, oxygen-depleted relatively warmer waters may have risen up out of the ocean basins, though the role that low oxygen waters upwelling onto the continental margins played at this time is debatable.

What is not debatable is that the rapid termination of glacial conditions delivered the second body blow to the Late Ordovician world, by taking the planet out of the frigid icehouse and returning it to the warm greenhouse over the course of 500,000 years or so. As a result sea levels rose rapidly, a strong thermocline developed and ocean deep waters, chemical weathering on the continents resumed delivering nutrients to nearshore marine settings, and the deep oceans returned to their dysareobic/anoxic state. Aside from the general character of environmental changes to marine habitats — with associated extinctions — this second transition entailed the post-glacial sea-level rise caused low oxygen waters to blanket shallow marine habitats on the continental shelves and continental margins with subsequent losses to the benthic faunas that had managed to 'hang on' through the previous low sea-level/cool episode.

Interestingly, several specialists in the palaeontology of the Ordovician extinction have noted that, among the cohort of species that survived the Ordovician glaciation, there appears to be a trend toward 'simplified morphologies'. In other words, species that were large or exhibited elaborate (sometimes bizarre) morphological structures seemed to have a greater likelihood of becoming extinct during one or the other of the two Ordovician extinction pulses than did species that exhibited small standard body sizes and more generalized morphologies. This distinction has been noted in the cases of Ordovician graptolites, bryozoans, conodonts and acritarchs. Drawing again on analogy with modern species, small generalized forms tend to exhibit shorter generation times and broader environmental tolerances.

One way to reduce body size and alter ecology is to shorten the time taken by a species to achieve reproductive maturity. If some individuals of a species mature reproductively while their bodies remain in a smaller, morphologically and ecologically generalized, juvenilized state, they may be better able to survive under adverse environmental conditions. This represents an evolutionary strategy that can result in lineages escaping the inherent limitations of morphological specialization that would be imposed on normal sized members of their populations. Patterns of evolutionary change in reproductive timing are known to have occurred in more modern species as a result of a population finding itself in a novel environmental situation (e.g. isolated on an island). As we will see, this somewhat unusual observation has also been made in studies of species survivorship across other ancient extinction events.

8 The Late Devonian extinctions

ollowing the end-ordovician extinction event the marine biota once more underwent another evolutionary radiation as new species colonized the extensive continental shelves and shallow marine seas, filling the new ecological roles created by rising sea level. The Silurian radiation differed from the Ordovician insofar as the old trilobite-dominated Cambrian evolutionary fauna, many members of which were still important components of the Ordovician marine realm, declined to a vestigial rump of species eking out a living in the margins of ecological structures dominated by the much more advanced group of species that palaeontologists have come to know as the Palaeozoic Evolutionary Fauna. This fauna included brachiopods with hinged, articulating shells (Articulata), bryozoans, the oddly structured stromatoporoid sponges, the coiled ammonoid cephalopods, advanced echinoderms (crinoids, blastoids, starfish) and leaf-like, colonial, floating graptolites. These groups made their first significant appearance in the Early and Middle Ordovician and underwent an explosive radiation in the Silurian and Early Devonian. In tandem with this marine invertebrate radiation, fish first become a significant component of marine and freshwater settings in the Silurian, land plants continued to diversify and animals (e.g. insects) began making their way onto the land (below).

OPPOSITE Sunset on the ancient Devonian reef cliff walls of Windjana Gorge, Kimberly Region, Western Australia.

Reconstructions of typical Devonian environments.

TOP Shallow water marine environment from the Canadian arctic. Foreground: the lobe-finned fish *Tiktaalik* (left and centre) which represents an evolutionary intermediate between fish and tetrapod amphibians and the early placoderm fish *Aterolepis* (right). Background: early actinopterygian fish, e.g. *Novogonatodus*.

BOTTOM Early terrestrial environment from Scotland. Foreground: lycopod, embryophyte, sporophyte, and trachyophyte plants including the genre *Asteroxylon* (left), *Aglaophyton* and *Horneophyton* (centre), *Rhynia* and *Nothia* (right).

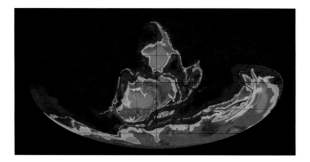
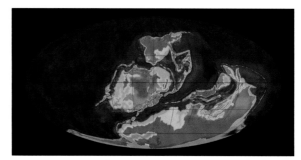

ABOVE Devonian palaeogeography. Left, Early Devonian (*c.* 400 million years ago), and right, Late Devonian (*c.* 370 million years ago).

SETTING

Geographically the old supercontinent of Gondwana was still extant and covered most of the Earth's southern hemisphere in the Silurian. Several much smaller continental fragments (Avalonia, Baltica, Laurentia) had drifted together near the equator. Over the course of the Silurian and into the Early Devonian these continental fragments would coalesce to form a new continent, Euamerica (above). The tectonic collisions that ensued as a result of this amalgamation formed the ancient Caledonian mountain chain of which the present-day northern Appalachian, Amorican, Bohemian and Scandinavian mountain ranges are remnants.

The large Ordovician ice-caps and continental glaciers receded during the Silurian as sea level rose. Climates were equable on the whole with relatively warm oceans and a small latitudinal temperature gradient. The interiors of the large continents were cold in the winter and hot in the summer, growing increasingly more so as the Devonian progressed. Stable isotopic data suggest that the composition of the Earth's atmosphere was relatively unstable at this time, with large excursions in atmospheric CO_2 occurring through the Silurian. These excursions appear to be associated with minor extinction-intensity peaks.

OPPOSITE Prominent victims of the end-Devonian extinction: trilobites (top left, *Phacops* from Morocco); foraminifera (top middle, *Endothyra* from the Brownwood Shale, Bridgeport, Texas, USA); rugose corals (top right, *Syringaxon* from the Wenlock Limestone, Malvern, England); conodonts (middle left, assemblage from the Timan-Pechora region of Russia); ammonoid cephalopod (middle right, *Gonioclymenia subarmata* from Shaldon Beach, Teignmouth, South Devon, England); sponges (bottom left, *Dictophyton tuberosum* from New York); and acritarchs (bottom right).

EXTINCTIONS

Despite the evident success of this Silurian–Devonian evolutionary radiation, the biosphere suffered another series of catastrophic extinctions in the Late Devonian, an event whose cause(s) are less well understood than those of the Late Cambrian and Late Ordovician. At this time over 20% of marine families and almost 60% of all marine genera vanish from the fossil record. Depending on taxonomic assumptions, this loss translates into between 80 and 90% of all marine species. These rates are comparable to those of the Late Ordovician extinction both in terms of percentage and absolute numbers of species lost.

Most groups commonly represented in the Devonian fossil record were affected by the Late Devonian extinctions. The exception was plants, which suffered remarkably few extinctions. Prominent victims included articulate brachiopods,

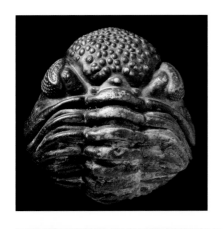

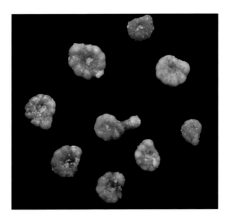

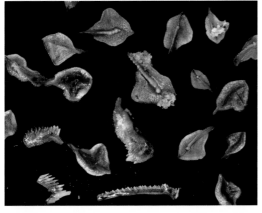

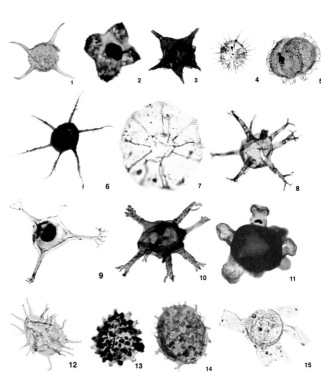

RIGHT *Dunklosteus*, a Devonian arthrodire, had jaws lined with bone plates. Also in the picture are sharks of the genus *Cladoselache*.

RIGHT The brachiopod genus *Mucrospirifer* shows, in dorsal view, its characteristically wing-like shell. This specimen is from the Devonian of Ohio, USA and is 3.5 cm (1½ in) wide.

corals, stromatoporoid sponges, trilobites, cephalopods, the scorpion-like eurypterid arthropods, marine fish, the engimatic, eel-like conodonts, marine plankton and the amoeba-like benthic foraminifera (see p.85). As with the Cambrian and end-Ordovician extinction events, the shallow tropical oceans were especially hard hit with the extensive Middle Devonian coral–stromatoporoid–bryozoan reef ecosystems collapsing to the extent where they effectively ceased to exist. Indications are that phytoplankton diversity – which forms the basis of most marine food chains – was also depressed to extremely low levels at this time.

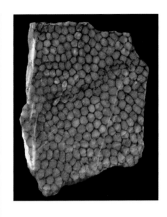

ABOVE The simple, polygonal corallites of the Devonian colonial coral *Favosites* are well seen in this 6 cm (2½ in) high polished block from the Devonian of southwest England.

TIMING

All researchers agree that understanding the time interval over which these extinctions took place is critical to understanding the cause(s) of this event. However, there is little agreement among specialists over this aspect of the Late Devonian extinction record. To a large extent this is a definitional problem. Some researchers prefer to separate out well-known regional extinctions that occur prior to the main extinction pulse (e.g. the Hunsrück, Kellwasser and Hangtenberg events), each of which is of lesser intensity and all of which have different possible causes. Under this scenario the end-Devonian extinctions occurred in two phases, an end-Frasnian event of relatively longer duration and greater overall intensity followed by an end-Famennian event that was much shorter (*c*. 1 million years). Other researchers see these events as component parts of a larger and longer-term interval of massive environmental deterioration. Overall the time interval during which these extinctions took place spans some 3 million years.

CAUSE(S)

Flowing from the complexity of its duration and timing, the set of cause(s) that have been proposed to account for the Late Devonian extinctions are many and varied. As with the end-Ordovician extinction event, differences between the Devonian world and our own need to be appreciated, particularly differences in the composition of the atmosphere. Oxygen composed less than 13% of the Devonian atmosphere, perhaps dropping to less than 10% in the Frasnian. This stands in contrast to its 21% abundance in today's atmosphere. The extremely low oxygen concentration in the Devonian atmosphere means that, for example, charcoal could not form. The greater proportion of greenhouse gases residing in the Devonian atmosphere also means that global temperatures were much warmer with a very wide belt of equatorial conditions, warm marine surface waters and very reduced equator-to-pole temperature gradients.

Various models of environmental change exist for the Late Frasnian events (including the Hunsrück, Kellwasser and Hangtenberg events) which emphasize

RIGHT Photograph of an outcrop of the Kellwasser black shale in this formation's type area in the Hartz Mountains, Germany.

RIGHT Photograph of an outcrop of the Kellwasser black shale in this formation's type area in the Hartz Mountains, Germany.

different aspects of the available evidence, but all involve a combination of sea-level rise, global cooling, and anoxia as the proximate extinction mechanisms. In one scenario (Buggisch 1991) rising sea level triggers deep water anoxia (evidenced by widespread Frasnian black shale deposition on the continents due to high primary productivity on the shallow Devonian seas). Anoxic bottom waters would progressively invade continental shelf habitats as sea-levels continued to rise causing benthic marine extinctions, while the draw-down of atmospheric CO_2 caused by high levels of marine productivity would reduce the concentration of this greenhouse gas thereby leading to global cooling and further extinctions, especially in the case of temperature and light-sensitive organisms such as reef corals. Many modern corals are physiologically dependent on the presence of the photosensitive dinoflagellate *Symbiodinium* in their tissues. If the same was true for Devonian corals, either the deepening of the water column above the reef or the filtering of light through a cloud-rich atmosphere would attenuate light penetration with prevailing cold conditions exacerbating a loss of physiological efficiency. In addition, corals consume bacteria, phytoplankton and zooplankton that are both ultimately dependent on sunlight. If submergence moved the available food source up away from the Late Devonian reefs, the reefs could have starved.

Taken to its logical conclusion this model predicts the formation of ice caps and eventual regression in sea-level thereby reversing these changes in a seemingingly mirror image to the end-Ordovician extinction. There is evidence to suggest the end-Frasnian extinctions coincide with a series of sharp cooling events. Isotopic data, indicates Frasnian sea surface temperatures were reduced by 5–10°C (8–18°F) on several occasions. These temperature drops are also associated with the deposition of black shales, signalling the introduction of anoxic water onto the Late Devonian continental shelves. Other evidence from this time consistent with the association of extinctions with global cooling includes the disappearance of reef ecosystems, the differential survival of high-latitude – presumably cold-adapted – species relative to tropical forms, and the proliferation of deep-water – again presumably cold-adapted – siliceous (= glass) sponges into what had been warm, shallow-water habitats. However, direct evidence for extensive Late Devonian continental ice-sheets is lacking at present. Sea level also stood high throughout the Late Frasnian suggesting that, if a Late Devonian ice-cap was present, it was small.

Alternative causal scenarios that invoke global cooling as a mechanism involve induction by a bolide impact event or an LIP volcanic eruption. Under these scenarios cooling is precipitated by large quantities of material being injected into the upper atmosphere increasing the Earth's albedo and causing an impact/volcano winter. Cloud cover generated by these mechanisms would also precipitate extinctions via the collapse of marine primary productivity.

However, there are difficulties with these scenarios as well. Bolide impact is a particular favourite of those who regard the Late Frasnian extinctions as having taken place very rapidly. While there is abundant evidence for bolide impacts in the Late Devonian in the form of impact sphereule deposits (see Keller 2005), no physical evidence that would confirm the occurrence of an extraterrestrial impact (e.g. geochemical anomalies of rare-earth elements, shocked quartz, microtektites) has ever been recovered from sediments associated with any of the Late Frasnian extinction horizons, much less the 'smoking gun' of an appropriately dated large impact crater. Absence of the latter is not surprising and certainly not definitive proof that a bolide impact could not have occurred. Owing to the sea-level highstand and the configuration of the deep-ocean basins at the time, it would be most likely that any single large bolide would land in the Devonian ocean. Since the entire Late Devonian sea floor disappeared down convergent plate boundaries literally hundreds of millions of years ago, the majority of physical traces of a large Late Devonian crater may have been removed from the Earth's surface. Nonetheless, the products of such an impact should have left traces in sediments deposited on the Late Devonian continents had such an impact taken place.

Aside from tectonically-induced sea-level rise and bolide impact, the other contender for the cause of the Late Frasnian extinctions is the Late Devonian Vilnuy Volcanic Province. Volcanic eruptions occurred in the eastern part of the Siberian platform and deposits from them are well exposed along the Vilnuy, Markha and Lena rivers (see p.90). New radioisotopic dates have located the Province's emplacement within the Late Frasnian–Late Famennian interval of elevated extinctions.

Geological mapping studies suggest more than 300,000 km^3 (186,000 cubic mi) of lava flows and ash beds were emplaced as a result of plate tectonic rifting that occurred between 340.0 and 380.0 million years ago. While more research on the extent and timing of this eruption is needed, data available at the time of writing indicate that the bulk of the eruption may have occurred between 360.3 and 370.0 million years ago. These dates for the most intense interval of activity are slightly early with respect to the Late Frasnian events, but it should be remembered they are preliminary. If more precise dating indicates the Vilnuy Province eruptions are associated with either the Late Frasnian extinction or Late Famennian extinction intervals it may provide a compelling cause for a set of specific extinctions. If not these eruptions will still remain a mechanism that may have induced environmental variability over the general Late Devonian interval.

RIGHT Location (inset), extent, and general structure of Vilnuy volcanic deposits in eastern Siberia.

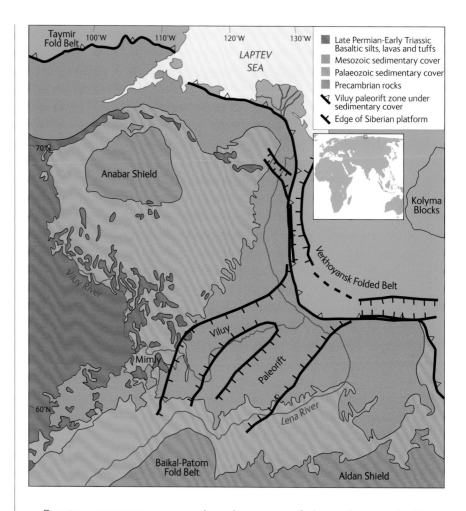

RIGHT Location (inset), extent, and general structure of Vilnuy volcanic deposits in eastern Siberia.

Despite uncertainties surrounding the causes of the earlier round of Late Devonian extinctions, a consensus does appear to have developed around the Late Famennian extinctions, which are widely regarded as having been caused by glacier-induced global cooling. Plate tectonic reconstructions based on Gondwanan palaeomagnetic data suggest this supercontinent moved to a position slightly north of the South Pole during the Late Silurian through the Middle Devonian, but reoccupied the South Pole during the latest Devonian. Drawing on analogy with the Late Ordovician extinction discussed in the last chapter, it has been argued that reoccupation of the southern polar region by Gondwana in the Late Devonian would have caused an ice-cap to form on this supercontinent accompanied by global refrigeration.

Late Famennian glacial sediments have been identified throughout Gondwana and alpine to lowland glaciers are known to have existed in eastern North America at this time. Precise dating of these sediments indicates that glaciation began as early as 364 million years ago, but reached its greatest intensity in the last million years of the Famennian. Tectonic forces responsible for the repositioning of Gondwana over the southern pole are assumed to be responsible for the onset of this intensive glacial episode, possibly in association with volcanic eruptions and palaeoceanographic factors.

The similarities between the Late Ordovician and Late Devonian extinctions are striking. Both involved the same evolutionary faunas, albeit in different proportions, both were of approximately the same magnitude, and for both the continental configurations were similar. Many of the ecological affects were also similar (e.g. loss of tropical reef systems, decimation of trilobites) and both resulted only in a set-back for the Palaeozoic biotas, which recovered from both extinction events and repopulated the oceans and, increasingly, the land surfaces. Yet the Late Ordovician and Late Devonian extinctions were also very different. The former coincided with a pronounced episode of continental glaciation and subsequent sea-level regression. The latter was not coincident with a sea-level rise and is associated with little direct evidence of global refrigeration except in its very latter – and comparatively minor – stages. It is hard to imagine that the processes responsible for both extinction events were the same. The Earth was simply in a different state during these two time intervals. Yet, the biotic effects that resulted were remarkably similar. The lesson I take away from this comparison is that similar biotic extinction effects can be produced by what would appear to be different causes operating in very different contexts and over very different time scales. While responses to stress can and do differ between species, in most cases an organism's response to proximal stress is the same regardless of the stress' ultimate cause.

9 The Late Permian extinctions

THE LATE DEVONIAN EXTINCTIONS eliminated most survivors and descendants of the old Cambrian evolutionary fauna. While a few representatives of this fauna would struggle on throughout the remainder of the Palaeozoic they were not merely subordinate components of shallow-water marine communities, they were genuinely rare. Other than this, the Devonian event had surprisingly little effect on the patterns of life at higher taxonomic levels. Naturally, a number of prominent constituents of Middle Devonian marine faunas did disappear from the fossil record in the Late Devonian (see pp.85-86). One interesting observation noted by George McGhee in a recent review of the Devonian extinctions is that all of the known Devonian lobe-finned fish species – close cousins of the common ancestor of all terrestrial vertebrates – became extinct in the Late Frasnian phase of that event. But like our own (unknown) Late Devonian ancestor, representatives of the vast majority of higher taxonomic groups survived the Late Devonian and spawned descendants who not only filled the ecological places of these species, but radiated into new and heretofore unsuspected Late Palaeozoic roles.

OPPOSITE Early Triassic sandstones below Runcorn Hill, Cheshire, England. While these sediments post-date the end-Permian extinction event they represent the same arid desert environments that characterized much of the lower and middle latitudes during the uppermost Permian.

SETTING

The brachiopods, corals, stromatoporoid sponges, cephalopods, eurypterid arthropods, marine fish, and even the enigmatic conodonts, the marine plankton, the benthic foraminifera and the trilobites, all came back after the Devonian in the form of new species with new variations on their group's body plans (see p.94). In addition, marine gastropod and bivalve lineages, which up to this time had been relatively minor members of marine faunas, began to diversify as the Devonian gave way to the Carboniferous and Permian periods. The most prominent exception to this pattern of broad-scale continuity was reefs. Members of the animal groups responsible for forming the frameworks of the impressive Devonian reefs (e.g. corals, bryozoans, sponges, calcareous algae) survived, but somehow the habit of coming together to form a robust physical framework that stood above the surrounding sea floor was lost for almost 100 million years, until a few modest reef frameworks were constructed by algae, sponges and bryozoans in the Late Permian. Among marine vertebrates the one immediately noticeable change was the disappearance of the heavily armoured fish — the placoderms — which were replaced ecologically by sharks.

Reconstructions of Carboniferous and Permian environments.

TOP Carboniferous marine environment from Montana, USA. Foreground: the coelacanth *Cardiosuctor* (top left), condrichthyans *Belantsea* (bottom left), *Falcatus* (top) *Stethocanthus* (middle) and *Harpagofututor* (right). Background: orthoconic and coiled ammonites.

UPPER MIDDLE Carboniferous terrestrial environment from the Czech Republic. Foreground: a palaeodictyopertidan (left) a dragonfly (centre) and the reptiliomorph amphibian *Gephyrostegus*. Background: the synapsid reptile *Archaeothyris* (middle) and the nectridian amphibian *Sauropleura* (middle, submerged).

LOWER MIDDLE A Permian marine environment showing trilobites.

BOTTOM Permian terrestrial environment from southern Africa. Foreground: the procolophnian parareptile *Owenetta* (lower left), the therapsid reptile *Diictodon* (centre left), the cynodont reptile *Procynosuchus* (centre right), the large gorgonopsian *Cyanosaurus* (top right). Background: the thereocephalian reptile *Moschorhinus* (left) and the dicynodont therapsid *Lystrosaurus*. (centre and right).

On land, the story was the largely same, but the character of the re-diversification was, if anything, even more profound. In the Late Ordovician and Silurian plants first moved out of the water and onto the land. By the Devonian the first forests had appeared. Also in the Devonian primitive insects had joined the marine exodus. Neither the plants nor the insects were affected by the Late Devonian extinction event. In the wake of that event, though, amphibians had not only appeared but begun to diversify, followed rapidly by appearance of the first reptiles. By the Late Permian advanced reptiles 3 metres (10 ft) long stalked the planet and had adopted a variety of new herbivorous and carnivorous life styles as their diverse – sometimes bizarre – morphological adaptations attest.

Geographically, Carboniferous Gondwana drifted more fully over the South Pole see below). This fact has led many to suspect that a permanent southern ice-cap was a feature of the Carboniferous world. In addition, the continuing continent–continent collision between northwestern Gondwana and Laurentia (consisting of Euamerica, Baltica and the Ural Terrane) raised the present-day southern Appalachian and Ouachita mountain ranges in the USA to alpine heights, probably changing the character of global atmospheric circulation patterns as well as the physiography of the land surface. By the Late Carboniferous Gondwana and Laurentia were sutured together to form a single supercontinent, Pangea, that stretched from pole to pole. A semi-enclosed sea was also present off the eastern coast of Pangea – the Tethys Ocean. Beyond this the Earth's surface was covered by a single, continuous, Panthalassic Ocean.

Sea level stood high throughout the Early Carboniferous (c.150–250 m above present-day sea level) when the Gondwanan–Laurentian collision was taking place, but declined precipitously in the Late Carboniferous presumably in response to continental glaciation caused by location of the old continent of Gondwana astride the southern pole. However, sea levels varied widely over the whole of the period as glaciers repeatedly advanced and retreated for reasons that are not well understood.

Climatically, the Carboniferous was characterized by well-developed cold, cool-temperate and tropical belts, the latter of which was quite wide, reaching above the 30th parallel in both hemispheres. Coincidence of the broad, shallow seaway that resulted from the Gondwanan–Laurentian collision with the tropical belt, and the evolution of the first woody plants – lycopods and conifers – resulted in the development of vast marginal marine and freshwater swamps in what is now the eastern USA,

BELOW Palaeogeography of the Late Carboniferous (left) c. 300 million years ago and Late Permian (right) c. 260 million years ago) worlds showing positions of continental landmasses and ocean basins.

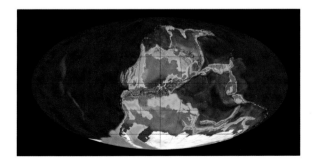
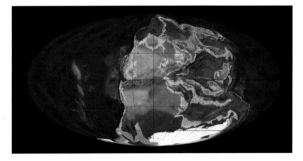

Britain, Germany, eastern Europe, Russia and China. The Late Carboniferous sea-level fluctuations alternately exposed and drowned these swamps, causing vast quantities of undecomposed organic material (largely plants) to be buried by new growth where they had once stood. This material was, over time, compressed and converted into the coal deposits that are found in these regions today. These Carboniferous swamps also served as the equivalent of modern-day biodiversity hotspots where the production of new marine and terrestrial species was enhanced.

With continued uplift along the tropically located Gondwanan–Laurentian suture zone, this shallow seaway dwindled during the Late Carboniferous, and finally disappeared along with the associated Carboniferous swamp tracts. In their place the alpine ranges of Late Palaeozoic mountains rose up in the middle of a (now) contiguous Pangean landmass. Without the ameliorating effect of a shallow seaway running through the middle of Pangea, the continentality effect caused global temperatures in the Pangean interior to rise. Over millions of years the broad Middle Carboniferous tropical belt was replaced by an even broader Late Carboniferous arid zone marked by extensive deserts. This dramatic change in climate regime is signalled not only by palaeogeographic reconstructions, but also by vast Permian-aged deposits of salt, gypsum, and other products of marine water evaporation that stratigraphically overlie the Carboniferous coal fields. Across Pangea arid conditions predominated almost to the 60th parallel, beyond which a narrow temperate band yielded quickly to very cold conditions in the planet's high latitudes. In effect, the Permian biota was caught in the merciless grip of cold climates at the poles and extreme hot and arid climates throughout the low and middle latitudes.

By the beginning of the Late Permian (p.95 bottom right) this trend towards extremely hot and dry, and cold conditions had reached the stage where vast numbers of marine and terrestrial species were at risk. Then something happened in the Late Permian that pushed the living world over the edge ecologically, triggering the largest extinction event ever recorded on Earth – referred to by some as the Great Permian Dying.

EXTINCTIONS

Palaeontologists as far back as Phillips appreciated that the end-Permian extinction was big. But the numbers David Raup and Jack Sepkoski estimated in the 1980s (see Table 3) are astounding even by geological standards: over 50% of all marine families, over 80% of all marine genera. The statistics for individual groups are even more grim: 98% of all crinoids and blastoid echinoderms, 96% of all anthrozoan corals, 80% of all brachiopods, 79% of all bryozoans. Figures for the terrestrial realm are equally stark: 77% of all land animals. Taken together these estimates suggest that well over 90% of all Late Permian species likely to leave a fossil record had vanished by the end of that period. Only terrestrial plants came through the Late Permian extinction largely unscathed, though the very high extinctions of marine filter-feeders (e.g. corals, bryozoans, crinoids, blastoids) suggest a severe reduction, if not a wholesale collapse, of marine primary productivity. Prominent victims of the Late Permian extinction event are shown opposite and p.98.

OPPOSITE Prominent victims of the end-Permian extinction: brachiopods (top left, *Cyclacantharia* showing this genus' solitary coral-like shape and long supporting spines, Glass Mountains, Texas, USA); gastropods (top centre, *Bellerophon bicarenus* from Tournai, Belgium), rostroconchs (top right, *Hippocardia herculean* from Tournai, Belgium); amphibians (middle left, *Eryops megalocephalus* from Geraldine, Texas); bryozoans (middle right, a 5 cm (2 in) long branch of *Hexagonella* from Australia), and fish (bottom, *Palaeoniscus freislebeni*, a fossil fish from Permian strata, Midderidge, Durham, England).

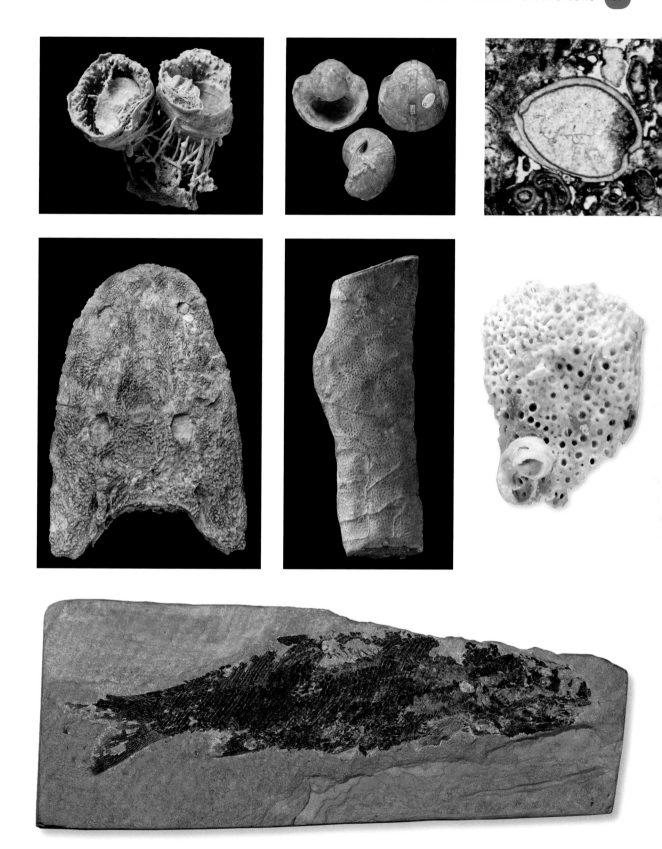

RIGHT Two sail-backed dimetrodons in the foreground with several edaphosaurs behind. Both of these genera were synapsid reptiles that lived during the Permian around 298 to 252 million years ago.

BELOW Prominent victims of the end-Permian extinction: cephalopods (top, *Goniatites crenistria* or *G. striata* from the Carboniferous limestone of Crowdicate, near Buxton, England); rugose corals (middle, *Lophophyllidium* sp.); foraminifera (bottom, fusulinid-rich biosparite limestone Stoner Member of the Stanton Formation, Kansas, USA).

These figures count all losses over an interval of at least several million years. But it needs to be remembered that all extinctions occurring in the Late Permian stages are counted in these summaries and that other species were evolving at the same time that this greatest of all extinctions was taking place. Also, the only species counted in these summaries are species that leave a fossil record. We know from looking at our modern world that most modern species will not leave any fossil record and that a number of ecologically important modern groups are virtually unknown as fossils.

While I have no doubt that many of the soft-bodied Late Permian species suffered the same fate as their hard-shelled or bony cousins, it is likely that the simpler, but no less important, bacterial, protist and fungal species came through this debacle in much better shape than the species that comprise the fossil record. This is, of course, a speculation, though one grounded on good data from investigations in the modern world. However, even if the last 5 or 6% of marine species left had succumbed to the Late Permian extinction event it is well to recall life evolved originally from simple bacterial beginnings in the Proterozoic under much harsher conditions than those that prevailed at any point in the Phanerozoic. Even if the evolutionary clock had been reset back to that point in the Late Permian, life would, in all likelihood, simply have picked itself up, dusted itself off and resumed evolving. Its resulting history would surely have taken different paths. But the point is, even though the Late Permian constituted a tremendous set-back for life on Earth, there is little evidence life itself was ever in any danger of going extinct.

TIMING

The interval of time over which the Late Permian extinction event took place is difficult to estimate because the magnitude of the sea-level regression that took place at this time means that most of the continental platforms were drained of seawater, much as they are today. During times of sea-level lowstand erosion scours down through the (then) recently deposited sediments. In most terrestrial and nearshore regions this process has removed all traces of the Late Permian rock record. The eroded sediment would ultimately have been redeposited in the ocean basins. But all ocean floor of this age has long since disappeared down the deep-ocean trenches as part of the plate tectonic process (see p.24). A near-complete record of latest Permian environmental and biological events exists in only a handful of places on Earth. While these windows on the past offer tantalizing glimpses of local sequences of events, the record they preserve is too restricted to offer a detailed understanding of the complete state of the Earth's biota across the entire globe.

What we do know is this. Elevated levels of species extinction begin to be noticed in the next to the last stage of the Permian, the Wuchiapingian. This pattern extended to, and was intensified in, the last stage of the Permian, the Changhsingian where most of the Late Permian marine extinctions are now believed to have occurred. [Note: I personally suspect that as research continues, as Wuchiapingian sediments are searched more intensively, and as new data are brought to bear on this extinction event (e.g. molecular systematics) a number of current extinctions assigned to the Changhsingian will be placed elsewhere in time.] Sticking with the evidence to hand, at a minimum the interval of time covered by these latest Permian extinctions is of the order of 5 million years. Even if we accept the current fossil record at face value, extinctions assigned to the Changhsingian cover an interval of 200,000 thousand years. In the most complete and best studied Chinese sections far more extinctions occur in the Changhsingian sediments below the Permian–Triassic boundary than at the boundary itself and more than 30% of the biota extant just below the Permian–Triassic boundary apparently survived into the overlying Triassic. This suggests extinctions occurred throughout the Changhsingian Age, not just during a short interval at its end.

Isotopic evidence suggests that coincident with the Permian–Triassic (P–Tr) boundary marine productivity declined sharply and global temperatures rose, indicating a short-term perturbation of the biosphere during the time interval that includes the extinction and the P–Tr boundary horizons. These same data also indicate a declining trend in productivity proxies, and a trend towards increase in global temperatures, which had been going on for hundreds of thousands of years in latest Permian. On a larger scale it is clear that Wuchiapingian–Changhsingian fusilinaceans, bryozoans and rugose corals progressively become restricted to the tropical Tethys Sea before disappearing from the fossil record.

CAUSE(S)

Many causes have been proposed to account for the Late Permian extinction. Some represent scientifically plausible scenarios that can be neither fully embraced, nor fully eliminated, because there is no evidence for or against them (e.g. irradiation by a nearby supernova). This set of potential causes must be set aside until some way is found to test their occurrence and timing. Putting these mechanisms aside, researchers are left with the usual suspects: tectonically induced sea-level change and temperature change (cooling or warming) induced by large-scale volcanism or bolide impact, or some combination of these processes.

Contrary to much that has been written about the Late Permian extinction, sea level change can be ruled out as a significant extinction mechanism. Sea level stood very low in the Late Permian, but this state was achieved at the transition between the Wuchiapingian-Changhsingian boundary which predates the bulk of the extinctions by several million years. The best current estimates of the character of sea-level change in the Changhsingian Age indicate that sea level was rising at a moderate rate throughout this entire time interval. Accordingly, none of the extinctions related to the last few million years of the Permian can be attributed uniquely to sea-level rise or any of its derivative effects (e.g. species-area effect).

Similarly, the vast bulk of Late Permian extinctions cannot be attributed to global cooling. As with the sea-level lowstand, the extensive Middle Permian glaciers had melted well before the Changhsingian began. Hallam and Wignall (1997) present a good case for extinctions in the Capitanian Age, leading up to the Capitanian-Wuchiapingian boundary, being induced by a combination of glacio-eustatic sea-level change and global cooling — including the famous Permian Glass Mountains reef complex in near Guadalupe, West Texas. But, again, these predate the bulk of the Late Permian extinctions by millions of years.

Global warming does appear to have been a dominant factor in the end-Permian extinction story. Evidence for globally warm conditions, except in the extreme high latitudes, comes from many sources including isotopic data (indicating a 6°C rise in the western Tethys over the course of the Changhsingian), palaeosol analysis (semi-arid soil types as far south as 60° latitude), and floral evidence. Moreover, the combination of low sea levels and globally warm conditions is signalled by the spread of Late Permian evaporite deposits. Indeed, so vast are the Late Permian salt deposits that some researchers have suggested that the salinity of the oceans was reduced, with consequent effects for organismal groups known or suspected to have a low tolerance for salinity fluctuations (e.g. rugose corals, bryozoans, ostracods).

While the extreme continentality of the Permian world would mean that high temperatures would have had a devastating effect on terrestrial plants and animals, by themselves they do not account for the widespread marine extinctions that occurred at this time. Many commentators believe that these marine extinctions were driven by 'a complex range of factors' which is probably true, but does not represent much of an explanation. Tony Hallam and Paul Wignall, however have consistently and convincingly argued that the real culprit was marine anoxia leading to a wholesale collapse in marine primary productivity.

Evidence for Late Permian marine anoxia is widespread in the form of the fine-grained sediments containing unoxidized carbon-rich minerals (e.g. pyrite) that characterize end-Permian sediments, preferential survival of low oxygen-tolerant species (e.g. ostracods, benthic foraminifera) as well as the extinction of low oxygen-intolerant species (e.g. reef species), and both sulphur and oxygen isotopes. Hoffman et al. (1991) have argued that the Permian oceans were structured in a similar manner to those of the Late Ordovician with anaerobic bottom water lying beneath an oxygenated surface layer. As with this earlier event, so long as the oceans remained unmixed this configuration is stable and organisms can adapt to it. However, nothing remains stable forever and any significant mixing would have disastrous consequences. If oxygenated surface water began to mix with deeper, anoxic water - say due to warming of the oceans to depth - not only could anoxic waters rise to the surface, but biolimiting nutrients such as phosphorus could be removed from the system over very short time scales leading to extinction of the phytoplankton and consequent destruction of marine food chains.

Moreover, the sea-level lowstand, coupled with hot, dry arid conditions in a broad tropical zone, would, in the absence of vigorous deep-sea circulation, promote deep-sea anoxia. During the subsequent sea-level rise these anoxic deep waters residing just offshore from the narrow Late Permian continental shelves would rise up to flood these shelves and go on to spread over the distal regions of the continental interiors, extinguishing countless marine invertebrate populations that had until then managed to survive through the species-area bottleneck that would have occurred as a result of the Late Permian sea-level fall.

Mechanistically how was Late Permian global warming and instability brought about? All of the complete Permian–Triassic sections have been examined closely for the tell-tale signs of an extraterrestrial impact. Retallack et al. (1998) claimed to have found a few shocked quartz grains of Late Permian age in deposits from Antarctica and Australia, but these were subsequently interpreted as plastic deformation structures more consistent with tectonic than extraterrestrial processes. Similarly, Becker et al. (2004) claimed that the Bedout structure, which is located in the sea bed off northwestern Australia, was the Late Permian analogue of the Chicxulub crater. But this interpretation, along with that of another structure in Wilkes Land, Antarctica, has not received support from independent investigators. While the hypothesis that a large bolide impact occurred sometime during the Late Permian has not — indeed cannot — be definitely disproven, evidence of the quality available for the association between a large bolide impact and the end-Cretaceous extinction event (see chapter 11) has yet to be produced.

In the absence of a credible argument for extraterrestrial forcing, the prime suspect is the well-known emplacement of the largest and most extensive volcanic field in Earth history that took place in Siberia during the Late Permian (see p.102).

These eruptions resulted in the extrusion of 2–3 million km^3 (1.8 million cubic mi) of basaltic lava that today cover almost 4 million km^2 (2.5 million square mi) to a depth of between 400 and 3,000 m (1,312 and 9,842 ft). Although for a long time it was thought that this lava field had been emplaced non-continuously over a period of tens of millions of years, new geochronologic dates now indicate that this entire package was emplaced in as little as 600,000 years, during a time interval that brackets the Permian–Triassic boundary.

The CO$_2$ and other gases released as a result of these eruptions would have changed the abundance of greenhouse gases with consequent effects on global surface and marine water temperatures. The range of possible environmental effects that would result from eruptions of this magnitude and over this timescale is shown opposite. Of course, the environment in the vicinity of the eruptive centres would have been devastated completely. But more globally, clouds thrown up by the eruptions would have reflected an increased proportion of solar radiation back into space, thus precipitating a short-term intensification of the longer-term tectonically driven global cooling trend. This intensification would be expected to result in some increased extinctions in both the marine and terrestrial spheres at or near the Permian–Triassic boundary, though this would be only one of a complex series of environmental effects. Naturally, this effect would also have been amplified by the Late Permian 'perched' state of the oceans.

While the links between the Late Permian extinctions and the Siberian Trap eruptions have been appreciated for some time, questions have been raised recently about whether even these events were sufficiently large to account for extinctions of the magnitude observed in the fossil record. Calculations suggest that lava flows consisting of thousands of cubic kilometres of molten rock would only have injected amounts of CO$_2$ into the atmosphere comparable to that released by modern society each year. Since such flows did not happen every year for millions of years, clearly an alternative source of greenhouse gases must have augmented the gases released as part of the eruptions themselves. Current thinking suggests these extra

RIGHT Location of the Siberian volcanic province on a reconstruction of Late Permian palaeogeography.

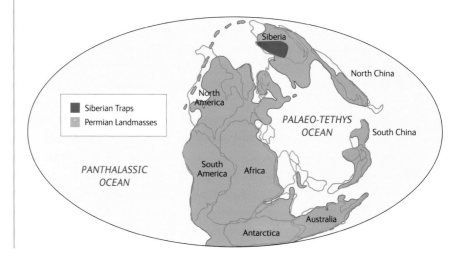

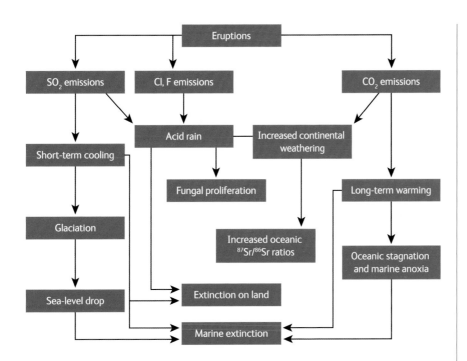

LEFT Diagram of primary environmental effects of the Siberian Igneous Province eruptions which occurred over a half-million year interval bracketing the Permian-Triassic boundary. (Redrawn from Benton 2003.)

greenhouse gases came from organic materials trapped in the sediments and rocks surrounding the Siberian Trap lava fields and (especially) from the uplift and exposure of Gondwanan coal deposits that occurred at this time. Mass balance calculations suggest these additional sources of CO_2 are more than sufficient to account for the Late Permian isotopic data.

It must be remembered that, in all likelihood Siberian Trap volcanism was merely one trigger – albeit a major trigger – that combined with the state of the planet at that time (i.e. a profound sea-level lowstand, extreme continentality, unprecedented globally high temperatures, marine deep-water stagnation) to cause the Late Permian environment to be in a profoundly unstable state. None of these factors was foreordained to have been operative and none was more important than the others in terms of contributing to the ensuing biodiversity debacle. Indeed the absence of any one would in all likelihood have resulted in a less intense episode of environmental change. But in the Late Permian a 'perfect storm' of environmental contingencies came together, like the end-Ordovician and end-Devonian extinction events, as a result of simple coincidence. The result was the largest single extinction event the Earth has witnessed to date.

10 The Late Triassic extinctions

I N THE EARLY TRIASSIC THE BIOSPHERE was populated by low-diversity, cosmopolitan taxa. This much has also been the case for each of the post-extinction intervals discussed previously. But like the Late Cambrian biomere extinctions, the end-Permian extinction event precipitated a fundamental change in the character of the Earth's biota. From the Early Ordovician to the end of the Permian life on Earth had been dominated by the Palaeozoic Evolutionary Fauna, an ecologically loose marine association of anthrozoan corals, articulate brachiopods, stenolaematan bryozoans, cephalopod molluscs, stellaroid and crinoid echinoderms, and the small, shrimp-like ostracods. Throughout this interval these groups comprised well over 50% of most fossil marine faunas. Yet across the Permian–Triassic boundary the number of genera assigned to these taxonomic groups halved. Through the Mesozoic their numbers averaged less than a third of marine assemblages. Today species belonging to these groups comprise just over 10% of the modern marine biota.

OPPOSITE The Triassic-Jurassic boundary section at Five Islands, Nova Scotia, Canada.

SETTING

Beginning in the Early Triassic the place of the Palaeozoic Evolutionary Fauna was taken by taxonomic groups we are more familiar with today, including calcareous nanoplankton, dinoflagellate plankton, foraminifera (benthic and planktonic), diatoms, sponges whose skeletons are made from spongin, silica or both (demosponges), bryozoans with an advanced muscle configuration for extruding their feeding organ (Gymnolaemata), snail (gastropod) and clam-like (bivalve) molluscs, the biscuit-shaped echinoid echinoderms, marine crustacean arthropods (shrimp, lobsters, krill), sharks (Chondrichthyes), marine reptiles and bony fish (Osteichthyes, see p.106). In the terrestrial realm insects, frogs, turtles, lizards, pterosaurs, archosaurs (including crocodiles, non-avian dinosaurs and birds) and mammals should be added to this list. Members of these groups exchanged places numerically with the members of the Palaeozoic Fauna in the sediments that immediately overlie the Permian–Triassic boundary, after which they begin a progressive march to their current state of overwhelming dominance, in excess of 90% of extant species.

With respect to plants, as we have seen with previous extinction events, changes in the flora across the Permian–Triassic boundary were far less marked than changes in the fauna. Here the so-called mesophytic flora persisted into the Early Triassic, including lycopods, conifers, cycads and ginkgoes. However, ferns first enter the fossil record in the Triassic and have been a significant floral component ever since.

While the biotic world of the Triassic was populated by organisms with familiar body plans, the physical world was still largely unrecognizable in terms of the present-day. Pangea remained intact, but began a northward drift that moved the fused Gondwanan continents of South America, Africa, Antarctica and India off the South Pole. In turn, Siberia and Kazakhstan moved over the North Pole. Deep-ocean trenches also formed along the northern and eastern margins of the Tethys Sea as it began to close in response to the northward drift and clockwise rotation of Pangea. This resulted in the formation and progressive amalgamation of island arc landmasses along the northern and eastern Tethyan margins. These accreting landmasses would eventually become incorporated into Turkey, Iran, Tibet, China, eastern Asia and Indonesia, but existed at this time as barely emergent islands or completely submerged oceanic plateaus. Perhaps more importantly in terms of understanding the Triassic Earth, the positions of these landmasses – created as a result of tectonic processes – would have affected oceanic circulation patterns both within the Tethys Sea and between the Tethys and Panthalassic Ocean, with the Tethys Basin becoming increasingly more isolated throughout the Triassic Period.

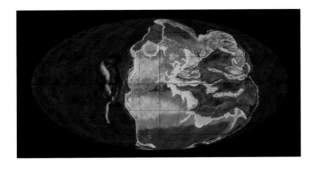

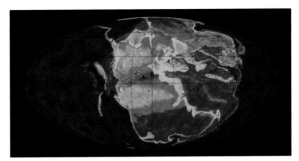

SEA-LEVEL CHANGES

During the Triassic sea level rose possibly in response to a reduction in levels of high latitude glaciation, but most likely as a result of tectonically-induced changes to the volume of the ocean basins. Some glacio-eustatic rise in sea level would be the expected effect of Pangea moving away from the South Pole, but there is little direct evidence for large ice sheets in middle-latitude settings in the Late Permian and no evidence for them in the Early Triassic. Present palaeogeographic reconstructions of Pangea in the Triassic suggest that broad marine embayments existed in the vicinity of the North Pole. These may have moderated the intensity of glacial activity in the northern hemisphere. Regardless, there is abundant evidence for the deposition of marine sediments on the Triassic continental platforms in Australia, India, Arabia and eastern North America, all of which were situated along the southern and western margins of the Tethys Sea at this time. Oceanic plate subduction and associated mountain-building activity characterized the western margins of North America, South America and Antarctica during the Triassic, keeping the continental shelves in these regions fairly narrow.

In the Late Triassic, however, this trend reversed itself. Throughout Europe and the North American Arctic there is abundant evidence for a substantial and geographically widespread sea-level fall. The lines of evidence for this regression are many and varied, including marine hiatuses with subareal erosional surfaces, exposure of reef tracks and pronounced shallowing of carbonate platform environments. Even more significantly, following this regressive phase marine deposition returns, but often in the form of dysaerobic/anaerobic black shales, signalling the spread of deep-ocean, oxygen-poor waters over the continental shelves. As we have seen before, this regressive-transgressive/dysaerobic couplet is often associated with widespread marine extinctions. The ultimate causal mechanism(s) for this sharp regressive–transgressive pulse are obscure, but this interval appears to be associated with the early phases of the tectonic rifting between North America, South America, Africa and Europe that would eventually produce the Atlantic Ocean.

Triassic climates were more equable relative to their very extreme states achieved in the Late Permian. Through the interval the area of extreme aridity – as defined by the locations of extensive evaporite deposits – shrank back towards the Triassic equator accompanied by the development of broad warm-temperate and tropical

ABOVE Palaeogeography of the Early Triassic (left) (c. 250 million years ago) and Late Triassic (right) (c. 220 million years ago) worlds showing positions of continental landmasses and ocean basins.

OPPOSITE BELOW Late Triassic terrestrial environment from Kyrgyzstan. Foreground: the beetle genera *Notocupoides* (lower left) and *Hadeocoleus* (centre left), the early gliding reptile *Sharovipteryx* (centre), the titanopteran insect genus *Gigatitan* (upper right). Background: the actinopterygian fish *Saurichthys* (top centre), the long-spined (or porto-feathered) reptile *Longisquama* (centre) and the synapsid reptile *Madygenia* (top centre).

regions in which coal swamps formed. Zones of coal formation were especially numerous in the northern hemisphere. Nevertheless, arid conditions, as signalled by calcrete and evaporite deposition, continued along the northwestern margin of the Tethys, in the areas now found along the western margins of Europe, northern Scandinavia and eastern Greenland.

Into this Triassic world evolution brought forth several interesting innovations in the history of life. Most important ecologically were the modern phytoplankton groups: calcareous nannoplankton and dinoflagellates. Throughout the Mesozoic these groups, along with the diatoms which radiated later, constitute the basis for marine food chains from the Mesozoic to the present day. As such they play critical roles not only in terms of marine ecology, but phytoplankton also sit – or more appropriately float – at the critical nexus between the physical and organic realms. These groups appear abruptly in the fossil record, suggesting that they may have had a long, but up to now invisible, Late Palaeozoic evolutionary history.

In shallow marine waters, brachiopods, bivalved molluscs and ammonites were quick to take evolutionary advantage of the depopulated ecological space. Each of these groups underwent significant adaptive radiations in the Early Triassic, radiations that were quite remarkable. Unlike the phytoplankton radiations which may have had a long pre-Triassic history, we know that all Triassic brachiopods and molluscs are descended from a very few Late Permian extinction survivors. Reefs also reappear in the Tethys by the Middle Triassic and grew to impressive proportions in many tropical and warm-temperate regions along the northern and eastern Tethyan coast. A significant addition to this biome was the appearance of modern corals, which today dominate most present-day marine reef ecosystems.

By far the most impressive Triassic evolutionary developments, however, took place among the vertebrates. Bony fish, sharks and rays all underwent radiations in the oceans, but had to share this ecological space increasingly with air-breathing interlopers. Pavement-toothed and seal-like placodont reptiles fed on the abundant bivalved molluscs and needle-toothed nothosaurs hunted fish in the shallow Triassic seas. The top marine predators of the Early Triassic were the fish-like ichthyosaurs. On land the mammal-like therapsid reptiles were an Early Triassic success story, but by the end of the period were replaced by the smaller, more gracile, and quicker thecodont reptiles including the thecodonts' most popular descendants, dinosaurs. Though dinosaurs first appear in the Triassic fossil record as tetrapods the size of small dogs, by the time of the Triassic–Jurassic boundary the lineage had produced giants over 6 m (20 ft) in length. True mammals also make their first appearance in the Triassic, but throughout the interval exist as rare, small, rodent-like species that fed on insects and/or scavenged for food. Despite the evident success of these new species of modern aspect, the good times of the Triassic were short-lived. A mere 25 million years after the cataclysmic Late Permian extinctions the Earth was again thrown into an environmental chaos from which only a much reduced fraction of marine and terrestrial species would emerge.

EXTINCTIONS

Curiously, while the end-Triassic is the second youngest of the 'Big Five' mass extinctions, it is the extinction earth scientists arguably know least about. To some extent this is the result of the lack of well-documented and complete Triassic–Jurassic boundary sections and cores, which, in turn, is a reflection of the effect of a major drop in global sea level across this boundary. However, as we have seen, all previous mass extinctions other than the Late Devonian also occurred at a time of a major sea-level lowstand. The other factor that conspires to complicate inferences is a relative lack of ways to accurately date Triassic sediments, at least compared with the dating methods available in overlying Jurassic, Cretaceous and Caenozoic intervals. Estimates of the magnitude of the end-Triassic extinction agree that over 20% of both marine and terrestrial animal families, and over 50% of then extant genera, leave the fossil record at this point. Simulations indicate that this corresponds to over 75% of all fossilizable species, which is roughly the same level of intensity as that recorded for the end-Devonian extinction event.

Once again reef ecosystems were decimated in the Late Triassic. Following the end-Permian extinction a 'reef gap' developed in the Early Triassic that was not filled until the Middle Triassic with the development of low mounds consisting of stromatoporoids, calcisponges and encrusting algae. Scleractinian corals appear in the Middle Triassic and were the dominant reef framework builders by the Norian, a time period which is referred to by reef palaeobiologists as the 'dawn of modern reefs'. However, this rise to dominance by scleractinian corals should be seen in the context of declining reef occurrence and diversity throughout the Middle and Late Triassic. Then, at the end of the Rhaetian Age the reef biome all but vanished, with over 95% of Rhaetian coral

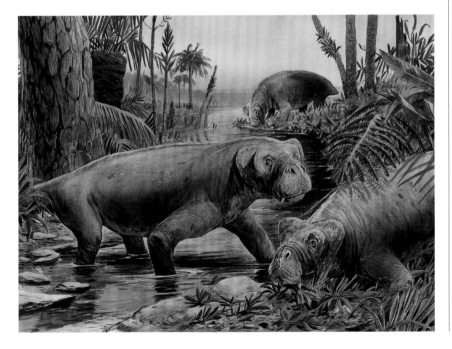

LEFT *Lystrosaurus* was a dicynodont, a mammal-like reptile that lived during the Early Triassic period c. 252-247 million years ago. Its fossils have been found in South Africa, India, Antarctica, China and Russia.

species leaving the fossil record. The last remnant Triassic reefs occur in the Tethys. Their disappearance meant that all reef-related carbonate deposition in the seaway ceased until a few, small reefs began to reform in the Early Jurassic. One potential implication of the disappearance of reefs is its possible association with a reduction in marine primary producers, a hypothesis that receives additional support from the extinction of zooplankton groups (e.g. Radiolaria). Bivalved molluscs were especially hard hit, losing over 90% of species. Brachiopods and ammonites suffered losses of over 80% and 60% of species respectively. These ammonite data are particularly noteworthy as they show clear evidence of a progressive extinction pattern with 150 Carnian ammonite genera being reduced to 90 in the Norian and just six or seven in the uppermost Triassic Rhaetian Stage. The ammonite record also shows evidence of differential selection against large, complexly coiled and (presumably) ecologically specialized forms. Among the microplankton, radiolarians suffered an extinction of around 69% of genera, but curiously little in the way of family-level extinctions. More noteworthy are the losses of whole organismal groups, the largest being the conodonts, the last representatives of which leave the fossil record entirely at this point, as do most labyrinthodont amphibians, mammal-like reptiles and many thecodont reptiles including a large number of dinosaur genera and species.

BELOW *Ichthyosaurus grendelius* with young, eating an ammonite. Ichthyosaurs lived during the Mesozoic Era, becoming extinct during the Cretaceous Period around 90 million years ago.

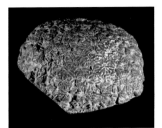

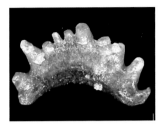

THIS PAGE Prominent victims of the end-Triassic extinction: ammonite cephalopod (left, *Ceratites nodosus* from Germany), corals (above top, *Astraemor* sp.), conodonts (above, *Hadrotontina* sp.) and nothosaurs (below, *Neusticosaurus pusillus* from Basano, Lombardy, Italy).

RIGHT AND BELOW Prominent victims of the end-Triassic extinction: right, placondont reptile, *Placochelys placodonta* from Veszprem, Bakony, Hungary, and below, amphibian, *Benthosuchus sushkini*, from the Scharzhenga river, Vachnevo, Russia.

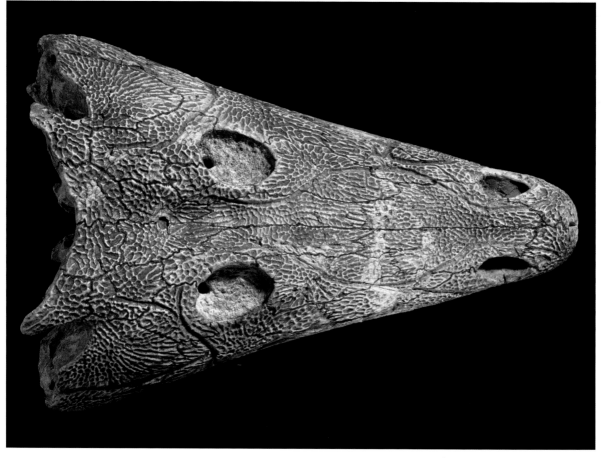

TIMING

Estimates differ on the timing of the Late Triassic extinction. Raup and Sepkoski's data – largely drawn from the marine fossil record – indicates the bulk of the extinctions were confined to the Norian Stage, whereas Mike Benton's summaries for terrestrial vertebrates suggest a two-stage event that also involves the previous stratigraphic stage, the Carnian. This latter interpretation has gained popularity and support from more recent studies of the marine record.

Lack of Late Triassic sediments accessible to palaeontologists has hampered our geographic understanding of this event. However, drawing on reefs as a model it seems reasonable to suppose that tropical regions were among the hardest hit. This speculation is not only in accord with reef distribution data, but also represents an expectation tied to the well-documented, and by some measures precipitous, fall in sea level that characterizes the Late Triassic. Marginal nearshore Tethyan habitats represent the last refuges for the Triassic reef communities. By the end of the period even these had disappeared from the fossil record.

Much controversy still surrounds the pattern of sea-level fall in the Late Triassic. Some geologists believe it was a protracted regression that began in the Middle Triassic. Others argue it was a more sharply demarcated event that was confined to the latest Triassic and continued into the Early Jurassic. Still others see unambiguous evidence for a sea-level drop in western Europe, but little evidence of a substantial decline elsewhere. Again, researchers are hampered by the dearth of latest Triassic sediments that could provide the data necessary to resolve this controversy.

Estimates of the magnitude of the sea-level regression also vary, ranging between 50 and 100 m (160 and 330 ft). These estimates identify the Late Triassic as being associated with the smallest sea-level fall of any of the major geological extinction intervals (except for the Devonian event which, you will recall, was associated with a sea-level rise). Regarding the causes of this drop in sea level, continental glaciation in the northern hemisphere resulting from the movement of northern Pangea over the northern polar region is an obvious suspect, though unambiguous Late Triassic glacial sediments – tillites – and/or dropstone deposits are rare. It also may be the case that the widespread proliferation of tectonic subduction centres through the interval may have contributed to sea-level instability through the creation of new ocean basins and variation in heat flow patterns at the mid-ocean ridges.

CAUSE(S)

Upper Norian sediments in Nova Scotia have been searched extensively for geochemical and physical evidence for a bolide impact coincident with the Triassic–Jurassic (Tr–J) boundary. Results of this search have been negative to date. In central Portugal a large (35-km or 22-mile diameter) circular topographic feature has been located, originally on satellite images, the so-called Guarda circular structure.

The origin of this structure has been dated very approximately to 200 million years ago. If future analysis shows this estimate to be accurate the formation of this structure would be coincident chronologically with the Tr–J boundary. Based largely on the evidence of this date and its circular structure this feature has been listed as a (possible) impact crater in some compilations. However, no unambiguous evidence for this interpretation has yet been published. In spite of this listing, most published summaries and/or reviews of the end-Triassic extinction event reject the idea of any link between the Late Triassic extinctions and an extraterrestrial impact.

While association of this extinction event with evidence for a bolide impact is highly tenuous at best, association with a huge volcanic province eruption is well established. In the Late Triassic exceedingly large igneous province eruptions took place along a front that extended across the entire (modern) eastern seaboard of the USA reaching as far west as Texas and possibly as far south as Venezuela. The basaltic lava units that comprise this province have counterparts of precisely the same age in northern Europe and western Africa. These eruptions represent the initial stages of the tectonic rifting of Pangea, known collectively as the Central Atlantic Magmatic Province (CAMP) (see below and Tanner 2004).

RIGHT Location of the CAMP volcanic province on a reconstruction of Late Triassic palaeogeography.

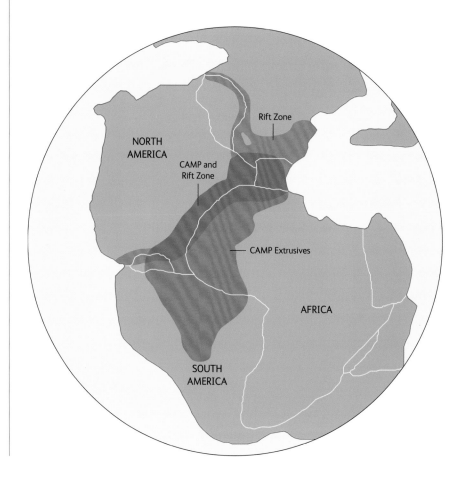

Overall the CAMP volcanics occupy a region of at least 10 million km^2 (6 million mi^2) with a volume of 2–3 million km^3 (1.2–1.9 million mi^3). This material was emplaced as a large series of separate eruptions that took place episodically along the entire rift zone. The total duration of CAMP eruptive activity is estimated to be about 500,000 years. Individual eruption centres were active for as long as 100,000 years, with most erupting for considerably less time.

In terms of timing, this event took place at 200 million years ago ± 2.0 million years. [Note: the aforementioned error estimate refers to the 200 million years radioisotopic date, not the duration of the entire CAMP emplacement event.] This date places the CAMP eruption in the Rhaetian, rather than the Norian Stage . Sepkoski's family and generic compendia did not include the Rhaetian as a Triassic stage, instead listing the Norian as the uppermost Triassic stage. Accordingly, association between CAMP dates and the timing of the Norian extinction peak remains viable under the reasonable assumption that Sepkoski lumped the Norian and Rhaetian stages together in his analysis.

Acknowledging the uncertainties associated with the correlation of palaeontological and geological data within this time interval, the extinction scenario that is consistent with the majority of the data is that the longer-term patterns we observe in the fossil record are probably the result of the Late Triassic sea-level fall. As we have seen with previous extinctions, falling sea levels would have reduced the extent of the warm shallow seas that occupied much of the Triassic continental platforms. Reduction of living space would bring marine organisms that could migrate to new locations into more intense competition with each other. Of course, sedentary species – including reefs – would simply die in place if environmental conditions exceeded species' tolerances. On land the same processes operate in reverse. The retreating coastlines would cause environments to migrate towards the ocean basins, forcing mobile species to move with them. But unlike marine communities that can rely on ocean water to maintain a constant environment over vast areas, the drained continents would experience greatly steepened latitudinal temperature gradients and increased environmental extremes due to continentality.

This mechanism matches the long-term patterns we see in the oceans. Details of the terrestrial Triassic fossil record are less well established, especially in terms of the timing of major biotic changes. But there seems nothing inconsistent with this general explanation for Late Triassic extinctions, especially when patterns of competition with the new sets of successful adaptations created by evolution are taken into consideration (e.g. the displacement of lystrosaurs by thecodonts). At the very end of the Triassic, eruption of the CAMP volcanics put further stress on the entire biosphere. This amplified global cooling patterns due to increased planetary albedo and changes in levels/patterns of erosion and nutrient supply as a result of greenhouse gases being injected into the atmosphere.

11 The End-Cretaceous extinctions

B
Y NOW THE STORY IS WELL KNOWN. Walter Alvarez, geologist son of the prominent physicist Luis Alvarez, was engaged in a research project to measure the age and rock accumulation rate in the Cretaceous–Palaeogene (K–Pg) boundary section near Gubbio, Italy, as part of an international effort to refine age determinations for major geological boundaries. Long having been intrigued with the end-Cretaceous extinction event, the younger Alvarez wanted to find out how rapidly the change between characteristically Cretaceous fossils and characteristically Paleogene fossils had taken, but was stuck for a method that could infer absolute ages in rocks so old to a resolution of thousands of years. He asked his father for advice. The elder Alvarez suggested a search for rare earth elements such as iridium, which are known to be raining down on the Earth from space at a constant rate.

Upon sampling the boundary clay, the Alvarez team, which eventually included the chemists Helen Michel and Frank Asaro, found a spike in the concentration of iridium at the boundary between the Cretaceous chalk and the Palaeogene clay – the K–Pg boundary (see p.118). The magnitude of this iridium anomaly was such that a simple reduction in sediment accumulation could not account for its presence. The only explanation the team could think of was that an anomalously large amount of iridium, and perhaps other rare earth elements, had been injected into the Earth's environment at a time coinciding with the K–Pg boundary. Only two potential sources for this material exist, the Earth's mantle and extraterrestrial objects, other planets. moons, comets and asteroids. Measurements of the amount of iridium leaking from modern volcanoes seemed to rule out a volcanogenic origin, which left bolide impact as the only reasonable source. Based on assessments of the amounts of iridium contained in known meteorites the size of the object was estimated at 10–12 km (6–7 mi.) in diameter, later reduced to between 4 and 6 km (2.5 and 6 mi.).

The 1980 announcement of the iridium anomaly's discovery and its interpretation, rocked geology to its foundations. Prior to this extraterrestrial impact was regarded as a 'far fetched' cause to invoke as an explanation for any aspect of the fossil record. As a matter of fact this mechanism had been proposed prior to 1980, but always as a logical possibility invoked in the absence of any supporting data. The difference this time was that the Alvarez team supplied convincing empirical evidence that such an impact had indeed occurred just a few tens of millions of years in Earth's past and, more important still, was associated with a large extinction event.

Evidence confirming the original discovery and its interpretation came in quickly, including the discovery of iridium anomalies in other (but not all) K–Pg boundary

OPPOSITE The Darting Minnow Creek, Texas, USA, K-Pg boundary locality which contains Chicxulub impact spherules near the base (Chicxulub spherule deposits are bottom right of the picture). Thin-bedded sandstones overlie the glauconite, phosphate and spherule-rich deposits. The K-Pg boundary is above the event deposit.

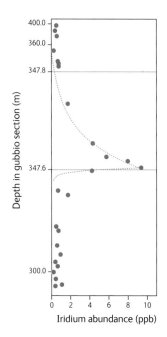

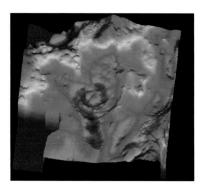

sections and cores, the discovery of other (albeit smaller) iridium anomalies associated with other (smaller) extinction events, and the discovery of other physical evidence of large K–Pg bolide impact, including quartz crystals whose structure had been deformed by the pressure of the impact (so-called shocked quartz, see above left) and round globules of molten glass that form when the heat and pressure of the impact eject molten rock out of the impact crater. By the late 1980s the case for a bolide impact having occurred at the K–Pg boundary was overwhelming. Then, in 1991 a team of geologists led by Alan Hildebrandt found what many believe to be the 'smoking gun' of the K–Pg impact, an unusual multi-ringed basin estimated to be *c.* 180 km (112 mi.) in diameter buried by Palaeogene sediments beneath Mexico's Yucatan Peninsula (see above right) – the Chicxulub crater.

For most members of the public, and indeed for the majority of the scientific community, it was 'case closed' for the end-Cretaceous extinction event. A giant rock from outer space had smashed into the Earth 65 million years ago and killed all the dinosaurs. How neat was that! Maybe giant rocks from outer space were the cause of other major extinctions in the fossil record. Maybe it was the cause of them all. The media loved it. The public loved it. But there was one group of sceptics who even today have never entirely bought into the asteroid-impact theory of extinctions: the palaeontologists – the people who study the fossils. Initial reactions of the palaeontological community have been described by David Raup (1991). For two decades after the 1980 announcement of the Alvarez team's discovery, international conferences were held to review new findings and update researchers from many fields on the status of developments across this interdisciplinary research programme. It became something of a tradition at these meetings to conduct an informal poll among conference delegates regarding the cause of the K–Pg extinction event. Those delegates who identified themselves as palaeontologists always returned the lowest levels of acceptance for the single-cause impact-extinction link in these polls. Why is that? Are palaeontologists just old-fashioned (in the words of Luis Alvarez) 'stamp collectors', jealous of the fact that scientists from outside their field had solved one of the classic palaeontological mysteries? Or is there more to the end-Cretaceous extinction story than many of those not familiar with the fossil record realize? To understand the answers to these questions we need to take a close look at the end-Cretaceous extinction event.

SETTING

For the most part species that survived the end-Triassic extinctions belong to the same groups that flourished in the Triassic. The conodonts were the only major group that disappeared completely from the fossil record by the end of that period. Nevertheless, the Jurassic and Cretaceous biotas are each quite distinct from those of the Triassic, an observation which owes as much to the appearance of new groups as to the diversification of old.

In the seas each of the major modern phytoplankton groups – nanoplankton, dinoflagellates and diatoms – either first appear or undergo their initial diversification in the Early Jurassic fossil record. It is something of a mystery as to what the dominant phytoplankton groups were in the Late Palaeozoic and Early Triassic. Judging by the diversity of Triassic reefs whose frameworks were composed dominantly of scleractinian corals that filter plankton from the water, phytoplankton surely existed. But whatever organismal group these species belonged to they have left no trace palaeontologists have yet been able to find. Irrespective of this mystery, the emergence of a new and stable basis for marine food webs revolutionized life in the sea. Not only did many existing marine groups diversify explosively at this time, but several important new groups appeared that made use of this new resource (e.g. planktonic foraminifera, deep-sea burrowing echinoderms).

The patterns that the Palaeozoic and Modern evolutionary faunas had established in the Triassic continued to be played out in the Jurassic and Cretaceous (see p.120). The Palaeozoic Fauna suffered a greater than 25% drop in generic richness across the Triassic–Jurassic boundary while the Modern Fauna recorded a greater than 25% richness increase in the same interval. Some Palaeozoic faunal groups increased in terms of numbers of genera through the Middle Jurassic, but then went into a protracted decline through the remainder of the Mesozoic (e.g. articulate brachiopods, crinoid echinoderms), whereas others continued to gain species at a modest rate throughout the interval (e.g. anthrozoan corals, bryozoans, ostracod arthropods). These gains were dwarfed, however, by the increases made by the Modern Fauna groups (e.g. gastropods, bivalves, echinoid echinoderms, crustaceans, bony fish). Ammonite diversification patterns were characteristically volatile throughout the interval where they were joined by a new, minor in terms of numbers but distinctive marine cephalopod group, the belemnites. Reefs reappeared in the Early Jurassic, their frameworks composed of an assemblage that included anthrozoan corals, coralline algae and bryozoans. In the Cretaceous, though, an unusual group of large, sessile bivalves – the rudists – also began forming accumulations large and tall enough to be considered reefs. Rudistid reefs became quite a common feature of the shallow Cretaceous seas from the middle part of the Cretaceous onward.

Changes were even more apparent, and dramatic, among Jurassic and Cretaceous vertebrate faunas. Ichthyosaurs survived the end-Triassic and re-diversified in the Jurassic oceans where they were joined by eel-like marine crocodiles and mosasaurs, as well as the morphologically eccentric plesiosaurs.

Reconstructions of Jurassic and Cretaceous environments.

TOP Jurassic marine environment from Switzerland. Foreground: the ichthyosaur *Stenopterygius* (upper left), the pliosaur *Rhomaleosaurus* (centre) with coiled ammonite and belemnite cephalopods (lower left) and the actinopterygian fish *Lepidotes* (lower centre). Background: the telosaurid crocodylian *Steneosaurus* (middle), the shark *Palaeospinax* (lower right), the crinoid echinoderms *Pentacrinus* (middle right) and the pterosaur *Dorygnathus* (upper right).

UPPER MIDDLE Jurassic terrestrial environment from the western US. Foreground: the dinosaur genera *Stegosaurus* (centre), *Diplodocus* (centre), *Allosaurus* (right) with the mammal genus *Frutafossa* (lower centre). Background: the dinosaur genera *Camtosaurus* (upper left) and *Apatosaurus* (top right).

LOWER MIDDLE Cretaceous marine environment from the Netherlands. Foreground: the mosasaur *Mosasaurus* (centre) a dead individual of which is being scavenged by the shark *Squalicorax* (centre), with belemnites (upper left) swimming past. Background: coiled ammonites (upper centre).

BOTTOM Cretaceous terrestrial environment from Alberta, Canada. Foreground: the dinosaurs *Chasmosaurus* (left), *Gogsaurus* (right) with *Lambeosaurus* and *Troodon* (centre right) with the metatherian mammal *Deltatheidium* (lower centre). Background: the orinthimimid dinosaur genus *Struthiomimus* (upper left) with the pachycephalosaurid dinosaur *Stegoceras* and the pterosaur *Quetzalcoatlus* (top left).

Both these groups enter the Jurassic fossil record as large animals and, by the Early Cretaceous, had fostered species of whale-like proportions. Jurassic and Cretaceous dinosaurs need little introduction here, as even young schoolchildren are conversant with their names, taxonomy and biology. Suffice it to say that the sauropods, ornithopods and theropods of the Jurassic (e.g. *Diplodocus*, *Iguanodon*, *Allosaurus*) and Cretaceous (*Alamosaurus*, *Triceratops*, *Tyrannosaurus*) were among the most impressive land animals of their own, or any other, time. But the Jurassic and Cretaceous were not just the heyday of the dinosaur. Birds (which in terms of their evolution and classification *are* theropod dinosaurs) also appeared in the Jurassic and diversified throughout the Jurassic–Cretaceous interval, as did pterosaurs. Arguably as significant – if not more so – a profound revolution in plant life also occurred during this interval with the appearance of the first angiosperms, or flowering plants, in strata of Jurassic age.

During the Jurassic and Cretaceous the Earth began to take on the character of the modern world physiographically (see below). Pangea stopped drifting northward as a unit and began to break apart in the Early Jurassic. Rifting between North America and Gondwana continued and, by the Late Jurassic, the central Atlantic Ocean Basin had opened to the extent that a broad seaway now separated North America and southern Pangea. From this point on Pangea no longer existed. To the north, North America became an island continent, like modern-day Australia. To the south, an amalgam of the present-day continents of South America, Africa, India, Australia and Antarctica drifted away from North America. This ancient supercontinent is usually, and somewhat confusingly, referred to by its old Palaeozoic name, Gondwana. The Tethyan Sea continued to close and a series of island arcs lying along the northern coast of the Tethys – formed as a result of tectonic subduction in that region – began the process of accretion into the emerging Asian landmass. In addition, by the Late Jurassic a broad, shallow seaway existed in the northern high latitudes with Europe and Scandinavia beginning to drift away from North America and Greenland.

By the end of the Cretaceous even more pronounced physiographic changes had taken place. Atlantic rifting continued to spread south and, by the end of the Cretaceous, Africa had separated from South America to the west and from the remainder of Gondwana to the south. Africa now also existed as an island continent. Similarly, India and Madagascar had broken away from Africa and existed as smaller

BELOW Paleogeography of the Late Jurassic (left) (*c.* 150 million years ago) and Late Cretaceous (right) (*c.* 90 million years ago) worlds showing positions of continental landmasses and ocean basins.

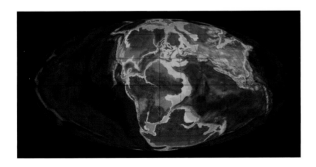

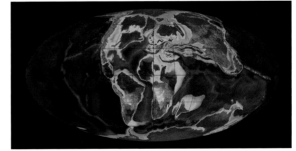

continental fragments in the proto-Indian Ocean. Australia and Antarctica were the only remnants of old Gondwana still together, but both had drifted south with Antarctica coming to lie over the South Pole once again. To the north the formerly deep and wide Tethyan Sea had all but disappeared as Africa began to collide with southern Europe and Asia. Most of western Europe was submerged beneath a shallow sea as a result of a sea-level highstand. A shallow seaway had also opened between North America and Greenland, while North America continued its northerly drift. The Late Cretaceous North Pole was covered by the shallow marine waters of a flooded Siberian continental platform.

In terms of climate, the largely warm (high-latitude) and arid (low-latitude) conditions that had predominated in the Late Triassic gave way to a more varied set of Jurassic environments. The equatorial arid zone marked by extensive Late Triassic evaporite deposits in North America, South America and Africa gave way to the development of middle latitude zones of warm temperate climates. This change is signalled by the occurrence of Early Jurassic coal deposits overlying Triassic evaporites in these regions. Palaeogeographic reconstructions suggest these warm, temperate belts occupied the zone between 30° and 60° N and S latitude. Above this a cool-temperate zone developed that had been absent completely in Late Triassic times. Presence of this zone is indicated by Early Triassic beds of iceberg and glacier dropstones in Siberia and Kazakhstan. Along the northeastern coast and hinterland of the Tethys Sea extensive tropical and subtropical climates predominated as evidenced by the widespread Jurassic coal deposits throughout China, Turkey, Iran and Tibet as well as in southern Europe (which was situated along the northwestern Tethyan coast at the time). However, by the Late Jurassic these tropical regions had changed into arid regions as evidenced by evaporite deposits that overlie the Chinese coals. This change is probably related to a short interval of sea-level regression at the end of the Jurassic.

During the Cretaceous the reorganization of global climates, begun in the Jurassic, developed further, aided by the organization of circum-global ocean surface current circulation in the northern high latitudes. An equatorial zone of tropical climates developed in the Early Cretaceous that promoted the preservation of extensive coal deposits in northern Africa, northern South America and southern Central America. This tropical warm and wet zone was bounded in the northern and southern hemispheres by an arid zone as evidenced by Cretaceous evaporite deposits in South America, Africa, North America and South China. This zone then passed into warm-temperate zones between 30° and 60° N and S latitudes that promoted coal formation, with cool-temperate environments above 60° N and S latitudes. Despite the large degree of climatic variation, however, the Late Cretaceous high latitudes were quite warm with both crocodilian and dinosaur faunas occurring above the Arctic Circle.

Over this Middle and Late Mesozoic interval sea level rose steadily as the climate warmed. Short-term sea-level regressions occurred in the Late Jurassic (Tithonian Stage) and the Middle Cretaceous (Cenomanian Stage). Interestingly these short-term reversals in global sea-level trends are associated with minor extinction-intensity peaks (see p.38). Some have argued that these extinctions are more regional than global in character. At the end of the Cretaceous though, sea level underwent quite a large and quite a rapid fall, dropping between 100 and 150 m (330 and 500 ft) in less than one million years. The magnitude and rapid onset of this regression remains a bit enigmatic causally insofar as there is little physical evidence of glacial activity sufficiently extensive to account for such a large drop in sea level in the uppermost Cretaceous stage (the Maastrichtian), though this lack of evidence could well reflect the fact that exposures containing the uppermost layers of Maastrichtian sediments are exceedingly rare.

EXTINCTIONS

More has been written about the end-Cretaceous extinction than any of the other Big Five extinction peak events. To some extent this is due to the popularity of dinosaurs, which are perceived (wrongly) to have become extinct at the boundary between the Cretaceous and Palaeogene time intervals. Nevertheless, the uppermost part of the uppermost stage of the Cretaceous (the Maastrichtian) through to the lowermost part of the lowermost stage of the Palaeogene (the Danian) interval is one of the most completely sampled and intensively studied intervals of all geological time. Unfortunately, the intense focus on the Maastrichtian–Danian boundary interval itself has often been achieved at the expense of larger, broader investigations, especially of the Late Cretaceous sediments that lie below the uppermost few metres of Maastrichtian sediments.

In terms of overall magnitude, the end-Cretaceous extinction is the least intense of the Big Five mass extinction events. On the basis of the loss of 16% of marine families and 47% of marine genera Raup and Sepkoski estimated the intensity of species extinctions to vary between 57 and 83%. While this percentage loss appears modest relative to those of the end-Permian and end-Ordovician extinction events, it should be remembered that overall taxonomic biodiversity increased exponentially throughout the Mesozoic. Over 2,400 genera have been found in the Maastrichtian fossil record compared to 1,239 for Sepkoski's richest Permian Stage, the Guadalupian, and 1,978 for his end-Ordovician Caradocian Stage. Accordingly, while the percentage extinction is decidedly less in the Maastrichtian, the total number of taxa lost is of the same order of magnitude as extinctions that appear much larger. The organismal groups traditionally associated with the end-Cretaceous extinction event are shown on pp.124 and 125.

Space limitations preclude a full discussion of the end-Cretaceous extinction records of all major organismal groups. For a more complete treatment interested readers are directed to MacLeod et al. (1997). Below I will only focus on the extinction records of dinosaurs, ammonites, planktonic foraminifera and terrestrial plants.

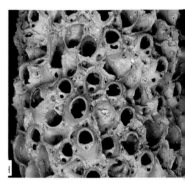

ABOVE Prominent victims of the end-Cretaceous extinction: plants as inferred from pollen grains (top, *Classopolis*, a type of conifer pollen); bryozoan (middle, *Melicerities squamata* from Denmark) and nannoplankton (bottom, scanning electron micrograph of a nannoplankton-rich Upper Cretaceous chalk,).

RIGHT AND BELOW Prominent victims of the end-Cretaceous extinction: ammonite cephalopods (top left, *Hoplites*, a strongly-ribbed Cretaceous ammonite from southern England); brachiopods (top right, a rhynchonellid brachiopod from Devon, England); inoceramid bivalves (middle left, *Cataceramus*, a subgenus of the bivalve *Inoceramus*); rudistid bivalves (far right); planktonic foraminifera (right, assemblage of Maastrichtian planktonic foraminifera from Brazos River, Texas, USA); pterosaurs (bottom, *Anhanguera blittersdorfii*) and cycads (below, *Bennettitales*, from Australia).

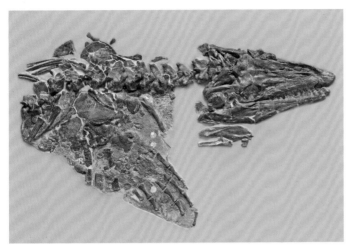

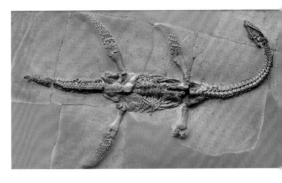

CLOCKWISE Prominent victims of the end-Cretaceous extinction: mosasaurs (top left, *Platycarpus* sp. from Elkander, Logun County, Kansas, USA); elasmosaurs (top right, *Plesiosaurus hawking*); metatherian mammals (above, *Zalambdolestes* from Mongolia); non-avian dinosaurs (left, *Triceratops*); ichthyosaurs (below, *Icthyosaurus communis* from Street, Somerset, England).

Ichthyosaur

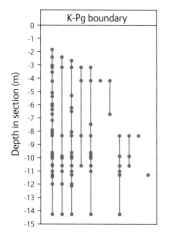

K-Pg boundary

ABOVE **Pattern of the occurrence of dinosaur fossil species (filled circles) in the Hell Creek Formation, Montana, USA. Horizontal line at the top is the level of the K-Pg boundary as marked by the local Ir anomaly.**

BELOW **The end-Cretaceous ammonite record at Zumaya, Spain (left). A detail of the uppermost Cretaceous ammonite record (right) with observed fossil ranges (red) and 95% confidence intervals on the last occurrences (blue).**

DINOSAURS

Dinosaurs are the single animal group nearly everyone associates with the end-Cretaceous extinctions. However, it should be noted that there presently exists only one window on the extinction record of latest Maastrichtian dinosaur faunas, the Hell Creek area of Wyoming, North Dakota, South Dakota and Montana. Based on this record, in the last 10 million years of the Cretaceous the number of dinosaur species was reduced by half, and over the last 15 m (50 ft) of the Hell Creek Formation only 11 dinosaur species have been discovered (left). This figure declines to three species in the uppermost 3 m (10 ft) of the formation with the last dinosaur remains occurring 1.77 m (5.80 ft) below the Maastrichtian–Danian boundary, which is marked by a local iridium anomaly. While the detailed chronology of this section remains to be determined, it is not unreasonable to suppose that the 1.77-m (6-foot) gap between the last non-avian dinosaur fossil recorded in the Hell Creek section and the level of the iridium anomaly represents at least thousands and perhaps tens of thousands of years. Moreover, the occurrence pattern of dinosaur remains in this succession is such that the potential existence of Palaeocene dinosaurs cannot be ruled out.

AMMONITES

Ammonites represent the invertebrate group most associated with the end-Cretaceous extinction. The Upper Maastrichtian ammonite fossil record has been studied most extensively in the Zumaya section in northern Spain. At a coarse level of stratigraphic resolution (below left) it appears as though 12 of the 30 ammonite species present in the Zumaya section disappear from the fossil record coincident with the Upper Maastrichtian stage boundary (= K–Pg boundary), an overall boundary extinction rate for this interval of 40%. However, focusing on the interval just prior to the K–Pg boundary in more detail (below right) it can be seen that none of these 12 species is actually recorded from the boundary horizon. If stratigraphic confidence intervals are placed on these ammonite ranges the extinction limits of most species fall well below the K–Pg boundary, though

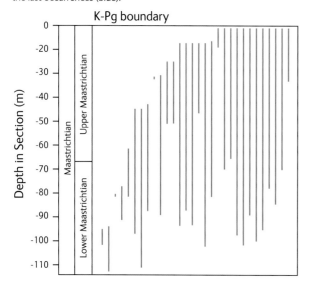

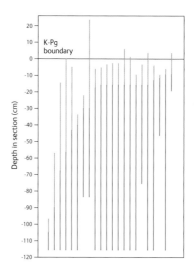

five species do have fossil occurrence patterns that suggest they may have reached the boundary, or indeed may even occur in lowermost Palaeogene sediments.

PLANKTONIC FORAMINIFERA

There has been much controversy surrounding the planktonic foraminiferal record across the K–Pg boundary. In many sections and cores the difference between the characteristically large-sized Cretaceous species and comparatively minute Danian species of this microfossil group is so striking you can see it with the naked eye. Reports of the first detailed studies of this group's K–Pg extinction pattern indicated that all but two or three Maastrichtian species existed at normal diversities right up to the boundary at which all but one or two species disappeared simultaneously (below left). But since 1988 our picture of the end-Cretaceous planktonic foraminiferal extinction has grown steadily more complex and well corroborated. In particular, Gerta Keller and her colleagues have documented a complex pattern of planktonic foraminiferal species turnover globally that involves extinctions prior to and at the K–Pg boundary as well as a sizeable and taxonomically consistent assemblage of 'Cretaceous' species that are found routinely in lowermost Palaeogene sediments (below right). Some of occurrences of some of these latter species' may be due to redeposition from eroded Late Maastrichtian sediments, a possibility that has always been admitted to by Keller and her colleagues. However, the consistency with which these Cretaceous species appear in lowermost Palaeogene sediments globally, along with their high abundance, excellent state of preservation and (in some cases) isotopic signatures all point to the majority of these fossils representing remains from living, Paleogene survivor populations. Recent independent estimates have suggested that as much as 30% of the Late Maastrichtian planktonic foraminifer fauna may have survived the end-Cretaceous extinction event.

Interestingly, these K–Pg survivor species tend to be small forms with simple morphologies that appear superficially similar to the morphologies exhibited by the fully Palaeogene species that first appear in lowermost Danian sediments. This occurrence pattern suggests the planktonic foraminiferal victims of the end-

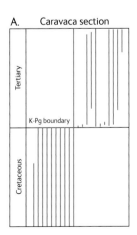

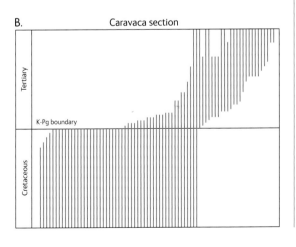

LEFT Two views of the end-Cretaceous planktonic foraminifera record at the Caravaca section in southern Spain. Left, an early interpretation that shows most species terminating at the Maastrichtian-Danian boundary horizon with only a single survivor species (Smit and ten Kate 1982) and a later interpretation documenting extinctions of 'Cretaceous' species prior to, at, and after the boundary horizon (Canudo *et al.* 1991).

Cretaceous extinction event were large, ornate, ecologically specialized and largely tropical species, whereas the survivors were small, morphologically simple, ecologically generalized, cosmopolitan species. The patterns of faunal turnover in this group also mirror similar patterns of turnover in the phytoplankton groups that form the basis of primary productivity in the marine realm. Unlike the wholly inferential interpretations of productivity collapse during the end-Permian and end-Triassic extinctions (where there is lack of a fossil record of phytoplankton species that the diversity of marine filter-feeders suggest must have existed), there is direct evidence for phytoplankton extinctions and abundance reductions in the uppermost Maastrichtian interval (see MacLeod *et al.* 1997).

PLANTS

Interpretation of the Maastrichtian–Danian plant record has undergone substantial revision over the past few years. In the 1980s and 1990s it was accepted that plants underwent a significant taxonomic turnover (*c.* 87% species extinctions) across the Maastrichtian–Danian boundary, with the largest proportion of the data coming from the American and Canadian west. However, studies published since 2000 have revised this figure down to about 50% based on megafloral analysis (e.g. leaves, fruit, flowers, seeds, cones) and as little as 15% based on analyses of pollen and spores. Both datasets are now recognized to exhibit the characteristic pattern of disappearances of species from the fossil record taking place before, at and after the Maastrichtian–Danian boundary horizon.

TIMING

Overall Jurassic and Cretaceous extinction records (see opposite) exhibit a striking shift in form. Extinction intensities in the Jurassic are, on the whole, higher and less regularly structured than those of the Cretaceous and Palaeogene. An initial radiation of species following the end-Triassic extinction took place through the Pliensbachian Age. After this interval average extinction intensities remained relatively high with the exception of the Aalenian Age, which represents an intriguing outlier with respect to the overall Jurassic pattern. The end-Jurassic Tithonian Age represents a minor extinction-intensity peak, probably related to a eustatic sea-level fall that may have been augmented by other factors (see Hallam and Wignall 1997).

In the Cretaceous, extinction intensities are, overall, much lower than those of the Jurassic, and organized into three distinct phases. A relatively brief interval of very low extinction intensity characterized the first three or four stages of the Cretaceous. This probably represents a post-Tithonian evolutionary diversification driven by sea-level rise in the oceans and the radiation of advanced terrestrial species (e.g. dinosaurs) on the land. Beginning in the Aptian, extinction intensities rise to a local peak in the Cenomanian. But rather than a sharp resetting of extinction pressures, whatever was happening in the environment at this time receded in intensity in a progressive

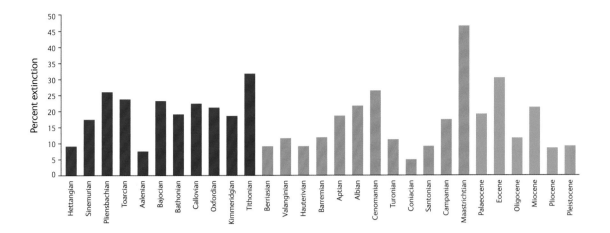

manner (on average) through the Turonian and Coniacian and then increased again spiking into the range of a major extinction event in the Maastrichtian. Extinction scenarios that seek to account for the end-Cretaceous extinction event must be consistent not only with the data observed in the Maastrichtian fossil record, but with the implications of this larger temporal context.

CAUSE(S)

As can be seen above, Cretaceous extinction intensities first began rising in the Aptian–Albian interval. Stable oxygen isotope data from planktonic (surface-water) and benthic (bottom-water) foraminifera indicate that, at this time, the Cretaceous seas began a long-term cooling trend (see below). Starting from a high of *c.* 20°C (68°F) in the Albian, marine surface and deep waters declined 3–5°C (5.4–9.0°F) to the Maastrichtian with pronounced deep-sea water temperature instabilities in the Late

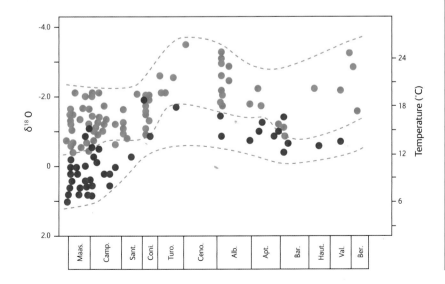

LEFT Cretaceous oxygen isotope history for marine surface planktonic (orange) and bottom benthic (purple) foraminifera indicating global temperatures of surface and bottom waters.

Campanian–Early Maastrichtian interval. There is abundant evidence from both fossil and modern organisms that water temperature decreases of this magnitude would be expected to place a number of marine groups under stress. Moreover, decreases in sea-surface temperatures of this magnitude would have global temperature repercussions, causing extinctions in terrestrial as well as marine lineages. Subsequent marine isotopic research has borne these inferences out and provided a more detailed picture of temperature-induced biotic stresses through the Late Cretaceous.

THE WESTERN INTERIOR SEAWAY

In addition, sea-level changes in the Cretaceous would be expected to be associated with increases in extinction intensity. The best record of the effect of the Late Cretaceous sea-level changes is preserved in North America (below). Here Euamerica was split into two continents (Larimidia to the west and Appalachia to the east) by a broad, shallow seaway known as the Western Interior Seaway. This seaway formed in

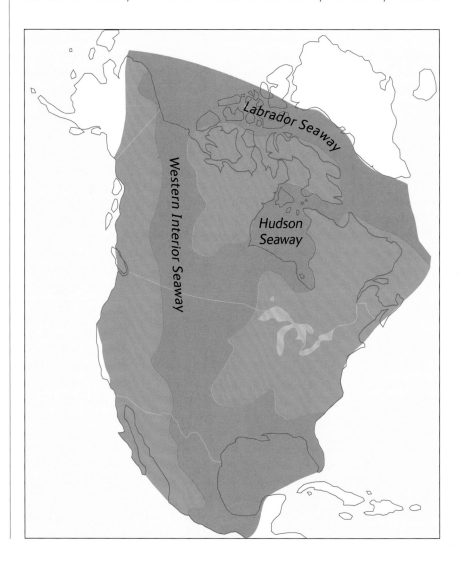

RIGHT The Cretaceous paleogeography of North America. See text for discussion.

the Early Cretaceous through a combination of the post-Tithonian sea-level rise and the depression of western Euamerica caused by the western margin of that continent drifting over the Farallon segment of a tectonic subduction centre that existed off the shore of western North America throughout the Jurassic. As the continental plate moved over this subduction centre the western margin of North America was pulled down as a result of the drag induced by the descending Pacific plate. This combination allowed marine waters to flood into the central part of Euamerica forming a broad, deep seaway that extended from the Gulf of Mexico in the south to the Arctic Sea in the north with intermittent connections to the Labrador Sea through a northeast–southwest-trending embayment that ran through the area occupied by the modern Canadian provinces of Manitoba and western Ontario – the Hudson Seaway.

The history of this seaway has two phases. The first lasted from the Albian to Middle Cenomanian Ages and led to the deposition of the Mowry Formation, which is a laterally extensive marine shale cropping out in the American states of Colorado, Utah, Wyoming and Idaho. This phase of the seaway – sometimes called the Mowry Sea – is well known for the ammonite and inoceramid bivalve faunas preserved in the Mowry Shale itself and for the Early Cretaceous dinosaur faunas preserved in time-equivalent terrestrial environments to the west. However, marine waters were drained from these areas during a sea-level fall that culminated in the Cenomanian. This caused the Western Interior Seaway to pull back to embayments in the area of the present Mississippi river mouth in the south and Alaskan North Slope in the north. As a result Larimidia and Appalachia were joined for a brief time in the middle part of the Cretaceous. This sea-level lowstand is marked on the extinction-intensity distribution (see p.129) by the Cenomanian extinction peak.

ABOVE An artist's depiction of the enigmatic drowning of a dinosaur in the Late Cretaceous Mowry Sea, based on the recovery of an actual fossil.

Marine waters returned to the Western Interior Seaway in the Late Cretaceous as evidenced by the chalk deposits of the Niobrara Formation and the deep-water shales of the Pierre Formation, both thick and widespread deposits of sedimentary rock that contain abundant marine fossils. It was during this Late Cretaceous interval that the seaway reached its maximum extent of some 1,000 km (620 mi.) in width and as much as 1,000 m (3,280 ft) in depth. These deposits are responsible for our clearest picture of what marine life was like in the Late Cretaceous. They even contain such exotica as pterosaurs and dinosaurs (see right).

Nevertheless, in the latest Maastrichtian, the Western Interior Seaway disappeared and did so within a remarkably short time. As with its creation, this disappearance was due to multiple factors. In the longer term, the rising Laramide mountain range – formed as a result of the same tectonic event that caused the seaway to be created – served as a source of sediments that flooded into the seaway from the west. Not only did this sediment fill in the western basin margins but its weight caused the middle of North America to subside further increasing the elevation of the land around the basin, increasing erosion rates and providing new accommodation space for the sediments shed as a result. Then, in the latest Maastrichtian global sea level fell by as much as 150–200 m (490–650 ft). The reasons for this sea-level fall are obscure as there is little evidence of continental glaciation in the Late Cretaceous.

Regardless of the cause of the end-Maastrichtian sea-level fall, the overall magnitude, the timing, or the effect of the sea-level fall are not in dispute. In less than one million years the Western Interior Seaway went from being one of the worlds largest epicontinental seas to being a few bumps in the coastline of a newly emergent, contiguous, North American continent. The sudden disappearance of the Western Interior Seaway in the Late Maastrichtian, including the marginal marine habitats that preserve our only record of latest Cretaceous dinosaurs, is coincident with the dramatic increase in extinction intensity observed over the Late Maastrichtian interval worldwide, and is correlative with the disappearance of a number of marine and terrestrial species that reached the middle part of the Maastrichtian in North America, but that disappear between this time and that of the K–Pg boundary.

DECCAN TRAP ERUPTIONS

In addition to environmental changes wrought by global cooling and sea-level fluctuations, the Late Maastrichtian Earth was rocked (literally) by intense volcanic activity. The geological map of India displays large red swatches in the central and western region (see opposite). These swatches represent areas covered by stacks of basaltic lava flows that form an enormous plateau covered with tropical vegetation and incised here and there with deep river valleys (see opposite top). These are the Deccan Traps, a large igneous volcanic province composed of the remains of an uncountable number of volcanic eruptions. The Deccan Traps outcrop pattern covers a surface of approximately 500,000 km^2 (300,000 mi^2), roughly the size of France. It must be realized that this is but a remnant of the volcanic deposit, essentially what was left after 65 million years of erosion. Moreover, the Traps extend out onto the continental shelf adjacent to western India. Indeed, submerged below the Indian Ocean a submarine mountain range made from basaltic lava flows, the Lakshadweep–Chagos Ridge, runs for about 4,000 km (2,500 mi.) in a north–south trend from a position close to the Central Indian mid-ocean ridge to the Deccan Traps (see p.134). This ridge forms the critical clue to the origin of the Deccan Traps and provides insight into the tectonic history of India.

Like the Siberian Traps in Russia (see p.102), the Deccan Traps were emplaced by a hotspot in the Earth's mantle. The basalts of the Lakshadweep–Chagos Ridge are all younger than the Deccan Traps and grow progressively younger as you move south along the ridge. This is the classic form of a hotspot track.

The Lakshadweep–Chagos Ridge terminates just north of the Central Indian Ridge, a tectonic spreading centre at which new ocean crust is created. Just on the southern side of the Central Indian Ridge, though, is another submarine accumulation of basaltic lava flows, the Mascarene Plateau, on the northwestern tip

LEFT The volcanic landscape produced by the Deccan Volcanic Province eruption that took place on the eastern side of the Indian subcontinent *c.* 65 million years ago and is associated with the end-Cretaceous extinction event. Note each light colored band of rock is a separate lava flow.

BELOW LEFT A geological map of India illustrating the extent of the Deccan volcanic field (red).

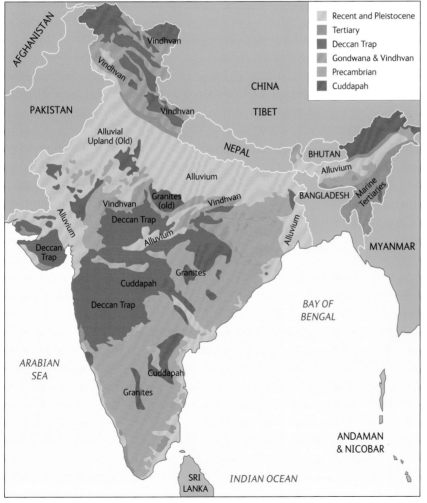

of which sit the Seychelles Islands. South of this, the Rodriguez Ridge terminates in the currently active volcanoes on the islands of Mauritius and Réunion. Dates for these basalts match the dates in the southern part of the Lakshadweep–Chagos Ridge, and then become younger as one moves south along the Rodriguez Ridge. In other words, the Deccan volcanic fields were created about 65 million years ago when the western region of the Indian subcontinent was positioned well to the south and west of its current location, over the Réunion hotspot. As India drifted northeast during the ensuing 50 million years, the lava outpourings from the Réunion hotspot created the Lakshadweep–Chagos and Rodriguez Ridges. The Central Indian Ridge is a comparatively new feature that bisected the hotspot track less than 40 million years ago.

For our story, the dates for emplacement of the Deccan Traps are fundamental. In 1981 Dewey McLean was the first to point out the correlation between the emplacement of Deccan Traps and the end-Cretaceous extinction event. At that time biostratigraphic

RIGHT AND BELOW Tectonic map of the modern and Cretaceous western Indian Ocean showing the link between the Deccan eruptions and the Réunion mantle plume. See text for discussion.

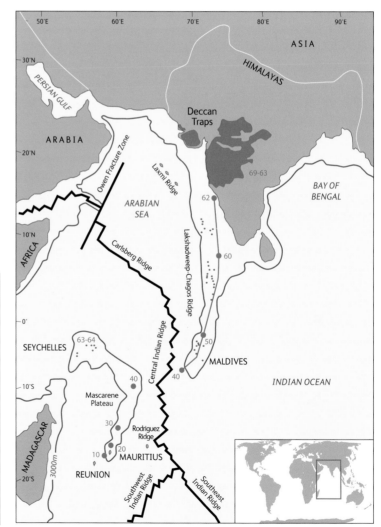

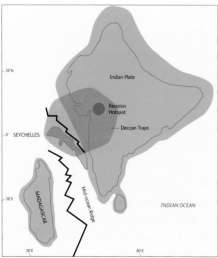

data coupled with palaeomagnetic evidence indicated these eruptions took place over an interval of not more than five million years at the end of the Cretaceous. These dates were circumstantially sufficient to link the Deccan eruptions generally to extinctions occurring in the Campanian and Maastrichtian, but not uniquely to extinctions occurring in the vicinity of the K–Pg boundary itself.

Subsequently, teams of geochronologists led by Vincent Courtillot have refined the Deccan Trap emplacement dates. What they have found is surprising. Courtillot's new dates demonstrate that the entire Deccan igneous province was emplaced in a little over one million years and during an interval that brackets the Maastrichtian–Danian boundary (see right). Even more recent work has reduced the emplacement time of the majority (80%) of the Deccan Volcanic Province to 50,000 years and demonstrated that significant planktonic foraminiferal extinctions took place over this entire interval (see Chenet *et al.* 2009 and Keller *et al.* 2011). The dating of individual lava flows indicates that, as expected, these eruptions waxed and waned in intensity over the interval. But the K–Pg boundary coincides with the most intense subinterval of Deccan eruptive activity. This coincidence is highly suggestive. Not only were the Deccan eruptions operative at the very end of the Cretaceous, they were at their most intense at precisely the time over which the bulk of the end-Cretaceous extinctions are known to have occurred. In geological terms this is as close to a 'smoking gun' association as earth scientists are likely to achieve.

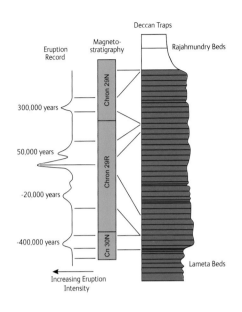

ABOVE Interpretation of the physical stratigraphy and geochronology of the Deccan Trap deposits.

Proximate extinction mechanisms

What was the extinction mechanism? There are several. As I have noted above, ash ejected into the atmosphere by volcanic eruptions increases the planetary albedo and can lead to global cooling. Carbon dioxide released by the eruptions themselves would have been augmented by CO_2 released as a result of baking the organic constituents of Palaeozoic and Mesozoic sediments over the entire Deccan region, both from above and below. Also, sulphur and chlorine, fluorine and related elements in compounds produced by volcanic eruptions break down in the atmosphere and combine with the water there to form various types of acids. These acids spread worldwide via atmospheric circulation and fell over the oceans and land as acidic rain. Based on recent atmospheric research we now known these same compounds can deplete the Earth's ozone layer that protects the surface from harmful ultraviolet radiation coming from the sun. Studies of recent volcanic eruptions – all of which are orders of magnitude smaller than the Deccan eruptions – show that tens of millions of cubic kilometres of this material can be released in very short periods of geological time in the form of ash clouds that can rise to heights of 50 km (30 mi.).

There is simply no question that historical eruptions have had major affects on the Earth's climate. In the wake of the Tambora eruption in 1815 average temperatures in the northern hemisphere dropped by 0.7°C (1.6°F) and as much as 10% of the ozone layer was destroyed. Extrapolating these data to the larger Toba eruption (69,000–77,000 years ago) suggests a global temperature drop of between 4 and 15°C (8 and 28°F) that lasted for decades and is believed by some to have initiated a miniature ice age that lasted for about a thousand years. Individual Deccan eruptions are believed to have been fissure eruptions, easily capable of lofting vast quantities of gas and dust into Earth's stratosphere. The similarly styled, but much smaller (14 km³, 8.5 mi³), Laki fissure eruptions that took place in Iceland from 1783 to 1784 released clouds of hydrofluoric and hydrochloric acid that killed over 50% of Iceland's livestock and induced a global famine that was responsible for the deaths of approximately six million people worldwide. Based on this evidence a wide cross-section of informed earth science researchers have concluded that formation of the Deccan Volcanic Province surely perturbed the global climate in (presently) unpredictable ways for more than a million years.

These considerations aside, there is another volcanically centred mechanism that could lead in principle to severe extinctions. First proposed by Dewey McLean, this mechanism is based on the reasonable assumption that vast quantities of CO_2 would be ejected by Deccan eruptions. In the modern world marine phytoplankton (especially nannoplankton) remove CO_2 from the atmosphere and store it as calcium carbonate ($CaCO_2$) in their skeletons. When these cells die their skeletons fall to the bottom of the ocean where they enter the stratigraphic record, eventually ending up as carbonate sedimentary rocks. To appreciate the level of activity of this process in the Cretaceous one need look no further than the name of the period: the root of the name 'Cretaceous' is '" which is Latin for 'chalk', one of the primary forms of carbonate sedimentary rock.

The Upper Cretaceous in particular is characterized by vast deposits of chalk stretching from Denmark (Møns Klint) through France and England (Dover Chalk), the southeastern USA (Selma Chalk), Texas (Austin Chalk) to the mid-continent (Niobrara Chalk). No chalks as laterally extensive and thick as the upper Cretaceous chalk deposits formed at any other time in Earth history. However, McLean and colleagues speculated that sufficient CO_2 would have been released during the Deccan eruptions to have disrupted this fundamental geochemical cycle. Essentially the oceans may not have been able to remove CO_2 at a rapid enough rate during the Deccan eruptions, causing this gas to accumulate in the atmosphere and surface waters of oceans, poisoning the phytoplankton due to acidification and shutting the CO_2 sequestration process down. Evidence for this comes from the sudden change from deposition of carbonate rocks to the deposition of non-carbonate silts and muds that marks the Maastrichtian–Danian boundary worldwide (see right), though the change in sedimentation style is also attributable, to some degree, to the extinction of the phytoplankton and zooplankton themselves for whatever reason. If it took place this effect would not be permanent. With cessation of the eruption (or at least its most intense phase) the system would have returned to its normal mode of operation. But during the disruption climate conditions on the Earth's surface would have been dramatically different. Primary production in the oceans would have been reduced to a very low value with consequent effects on marine and terrestrial food webs, acidification of marine and terrestrial waters, and significant global warming as a result of increased atmospheric CO_2 and destruction of the Earth's ozone layer.

Modern estimates of the effect of the shutting down of CO_2 sequestration scaled to estimates of CO_2 input that would accompany the Deccan Traps indicate these eruptions may have increased the temperature of the lower atmosphere by as much as 10°C (18°F), warming the oceans by as much as 4°C (7.2°F). If such estimates are even approximately correct many of the patterns and the scale of both biotic (e.g. extended interval of extinction activity, focus on tropical biomes, bias towards large-sized species) and abiotic data (e.g. presence and extent of K–Pg boundary clay, interval of wildly fluctuating Late Maastrichtian oxygen isotopic values, long interval of reduced carbon isotopic values in the Danian) collected from the K–Pg boundary interval make sense, especially in light of the accompanying sea-level regression.

THE CHICXULUB IMPACT

Despite the intriguing correlations and mechanisms associated with Deccan volcanism, there is yet another causal mechanism that was also operating in the environment precisely at the time of the Maastrichtian–Danian boundary – an anomalously large bolide impact. The strength of the case identifying the Chicxulub structure as an impact crater has already been discussed (see p.117–118). However, the precise timing of the Chicxulub impact remains a matter of dispute as well as whether it was the only Late Maastrichtian impact to have occurred (see Keller *et al.* 2004). Regardless, the simple demonstration of an impact's occurrence is not sufficient to identify it as the unique cause of an extinction event, especially for the end-Cretaceous extinction when so many other factors with equally strong claims to extinction causality are known to have been operating in the Earth's environment. The fact is, we now know that many sizeable impacts have occurred during the last 500 million years of Earth history, yet the only impact that has been causally associated with an extinction event is Chicxulub. A match between the effects of the Chicxulub impact and the observations made at the Maastrichtian–Danian boundary is required to evaluate the impact–extinction link properly.

LEFT Photograph of the Maastrichtian-Danian boundary in the section at Gubbio, Italy showing the end-Cretaceous boundary clay.

Proximate extinction mechanisms

In the original 1980 article announcing the iridium anomaly the Alvarezes, Asaro and Michel argued that one primary effect of the impact would be to plunge the Earth into and extended period of darkness accompanied by an 'impact winter' as a result of material blasted out of the crater and into the atmosphere where it would raise the planetary albedo. However, in 2002 geologist Kevin Pope argued that even a 10 km (6.2 mi.) impactor would have insufficient explosive power to cause a significant impact winter. [Note: the idea of an 'impact winter' originally came from US defense department simulations of the effect a nuclear war would have on the Earth's climate. The global cooling that was predicted to follow the explosion of hundreds of nuclear weapons across the northern hemisphere was termed a 'nuclear winter'.]

Acid rain was also an important kill mechanism alluded to in the original 1980 report. As volcanoes release nitrogen, chlorine, fluorine and other compounds that enhance the acidification of seawater, it was assumed that the Chicxulub impact would do the same. Given that the Yucatan Peninsula was covered in thick layers of carbonate and gypsum (both of which could serve as the source of acidic compounds) at the time of the impact, pH levels of 1.0 to 1.5 were predicted with confidence – about the same level of corrosivity as the acid in your car battery, which is made of thick plastic in order to contain the caustic materials it holds. However, the record of Maastrichtian–Danian species survivorship does not support this speculation. Development and industry also release such compounds into the atmosphere and, as a result, rainwater in certain areas has become more acid over the last 50 years than in the previous five million years. Accordingly we now know much about the effects of acid rain as well as the susceptibilities of different organismal groups. Based on these data it is difficult to understand how any terrestrial fish, amphibians, reptiles and a large variety of plant species could have survived long periods of exposure to rain as highly acid as was predicted by the original impact scenario. To be fair, McLean's volcanism scenario also calls for the production of acid rain. But these two scenarios differ in their predictions of the level of acidity of the rain and the duration of time it would be a regular feature of the end-Cretaceous and/or Early Palaeogene environment.

Finally, thermal radiation has been proposed by a number of authors and under a number of mechanistic guises (e.g. direct radiation from the impact itself, reflected radiation from low-lying clouds, wildfires begun when a rain of molten crater ejecta landed on a dry landscape) as accounting for large numbers of extinctions. Like many of the scenarios mentioned above, the radiation/fire mechanism is usually posited on physical grounds with little detailed explanation of which organisms would be differentially at risk under the scenario and no attempt to test the scenario using actual data. As summarized in a recent review (Archibald 2011), this mechanism too has not fared well either on theoretical grounds or in terms of explaining the details of actual Cretaceous extinction and survivorship patterns.

SYNTHESIS

Let me close this discussion by stating there is no question that at least one (possibly more) large impact(s) occurred during latest Maastrichtian time. There is also no question that this (these) impacts would have been devastating for local biotas and may have had short-term (less than 100 years) environmental consequences for the planet. But advocating bolide impact as the only credible cause for the end-Cretaceous extinctions is far too much of an oversimplification for myself and many of my palaeontological colleagues to accept, especially insofar as the mechanisms associated with bolide impact do such a poor job explaining the patterns we see in our palaeontological and geological data. [See Archibald 2010, a letter co-authored by 28 palaeontologists from across the palaeontological subdisciplines outlining opposition to the K–Pg impact-extinction scenario and written in response to an article published by Schulte 2010 supporting this model, but coauthored by only 12 palaeontologists the overwhelming majority of whom were microfossil specialists.] Based on a consideration of all the evidence presented above I find the causal framework that accounts for the greatest proportion of evidence is one that acknowledges the totality of physical events that are known to have occurred at the end of the Cretaceous: a scenario that invokes sea-level change, Deccan volcanism and the Chicxulub impact as all playing roles in precipitating the end-Cretaceous extinction event with terrestrial extinction deriving primarily from habitat fragmentation resulting from the end-Maastrichtian sea-level regression and marine extinctions driven primarily by a short-term collapse in primary productivity.

12 The Palaeogene extinctions

T HE EXTINCTION PEAKS THAT OCCUR in the Caenozoic – during the Late Eocene–Oligocene, Miocene, Pliocene and Pleistocene – are decidedly smaller than those of the 'Big Five' extinctions of the Palaeozoic and Mesozoic. Indeed, at present none are regarded as true 'mass extinctions'. But not very long ago a 'Late Eocene extinction' was grouped with the 'Big Five' in terms of possible cause and the Middle Miocene extinction peak was part of Dave Raup and Jack Sepkoski's evidence for extinction periodicity. Irrespective, of the status of these theory-based concepts, the Palaeogene and the Neogene extinctions are disproportionately important for understanding the patterns of species occurrence and ecology we see in the modern world as well as providing critical parts of the overall context within which we must view the modern-day biodiversity crisis.

Given the magnitude of the end-Cretaceous extinction event and the enormous amount of attention that has been paid to that, it is a bit surprising to note here that a relatively small number of higher taxonomic groups disappeared from the fossil record during the Maastrichtian Age. The most prominent victims of this event were marine molluscs — the ammonoids, and the rudist and inoceramid bivalves — as well as the marine reptiles (mosasaurs, plesiosaurs), aerial reptiles (pterosaurs) and one group of terrestrial dinosaur (the ornithischians). All other major groups, including the saurischian dinosaurs, which survive to the present day as birds, are represented in the Early Caenozoic by direct descendants of Cretaceous species. To be sure, much extinction occurred at the species and generic levels in the uppermost Cretaceous and lowermost Danian stages. But as far as the diversity of major organismal body plans goes, true losses were surprisingly few.

SETTING

Most surviving Cretaceous groups underwent sustained evolutionary diversifications from the Palaeocene through the Early Eocene as they filled ecological roles vacated by the Late Cretaceous species extinctions. Particularly noteworthy radiations include those of the diatoms, planktonic foraminifera, gymnolaemate bryozoans, malacostracan crustaceans, echinoid echinoderms, bony fish (Osteichthyes), mammals and birds. These groups are all members of the Modern Evolutionary Fauna, which underwent an exponential pattern of richness increase at this time. Notably, the remnants of the older Palaeozoic Evolutionary Fauna are characterized by steady-state overall family and

generic richness patterns during this interval, with some groups maintaining their numbers in the face of competition from more modern species (e.g. the stenolaemate bryozoans, articulate brachiopods, crinoids, starfish) and others declining in the face of competition that was amplified by patterns of end-Cretaceous extinction and survivorship (e.g. cephalopods). The only prominent exception to these trends was the corals (Anthrozoa), which had already been displaced from their reef-forming roles in the Cretaceous Period by the rudist bivalves and underwent only a modest diversification in the Palaeocene, possibly as a result of the persistence of cold-water conditions on the continental shelves during that series' early and middle stages (see below).

Geographically the continents were beginning to take up positions close to their current configuration. Throughout the Palaeogene Africa, North America, South America, Europe and Asia continued to drift northward, away from Antarctica which remained over the South Pole. In the Palaeocene, Australia was still attached to the northeastern portion of Antarctica, but rifting was well developed. By the Middle Eocene separation was complete and a seaway – the Tasmanian Gateway – had opened up between these two island continents. To the west, the southern tip of South America remained attached to Antarctica by a broad continental shelf. This effectively blocked the establishment of circum-Antarctic currents and so helped keep Antarctica warmer than it would have been otherwise. The opening of a deep-ocean passage between South America and Antarctica – the Drake Passage – in the Late Eocene and Oligocene would allow the modern southern ocean surface current system to become established with dramatic consequences for the Earth's climate and for the course of evolution.

The Atlantic Ocean Basin was well established in the northern and southern hemispheres during this interval and separation between the Americas and Africa–Europe–Asia increased. The Tethys was but a remnant of its Mesozoic counterpart with much of southern Europe, the Middle East and western Asia submerged beneath a shallow sea. India continued to exist as an island continent, but the initial stages of the continent–continent collision that would create the Himalayan mountain range were already starting to be felt along the southern coast of Asia. Generally speaking, at this time southern Europe and western Asia existed as a complex assemblage of small continental fragments that were beginning to accrete onto the margins of the European and Asian plates. This process continued throughout the Palaeocene and Eocene. The North Pole lay beneath a broad, shallow and largely ice-free sea.

Sea level was highly variable throughout the Caenozoic but exhibited a long-term declining trend across the entire interval. This variability is probably an intrinsic feature of the sea-level record. Earth scientists have devoted more time to studying Caenozoic sea-level changes due to the amount of Caenozoic sedimentary rock available for study combined with the incentives provided by commercial petroleum exploration. The fact is, the location and extent of petroleum reservoirs is, to a large extent, controlled by the distribution of packages of sediments that migrate in response to changes in sea level.

Reconstructions of Palaeogene marine and terrestrial environments.

TOP A Middle Eocene marine environment from Italy, including the actinopterygian genera *Exila* (upper left), *Lophius* (left), *Echolocentrum* (centre), *Psettopsis* (top centre), *Mene* (top right), the eel *Eomyrophis* (centre) and the ray *Tyron* (right).

MIDDLE An Eocene terrestrial environment from the western USA. Foreground: the condylarth *Hypossodus* (extreme left), the adapiform primate *Cantius* (left), the early artiodacyl *Diacodexis* (lower) and the condylarth *Phenacodus* (right).

BOTTOM Foreground: the crocodylian *Boraleosuchus* (left), the anseriform bird *Presbyornis* (middle), and the rhinoceros-like ungulate *Uintatherium* (right). Background: the creodont *Arfia* (extreme left), the early perissodactyl *Hyracotherium* (centre) and the bats *Icononycteris* and *Onychonycteris*.

Palaeogeography of the Palaeocene (above, *c.* 65 million years ago) and Eocene (below, *c.* 50 million years ago) worlds showing positions of continental landmasses and ocean basins.

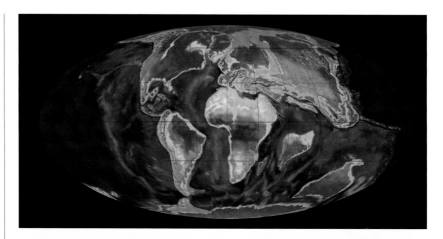

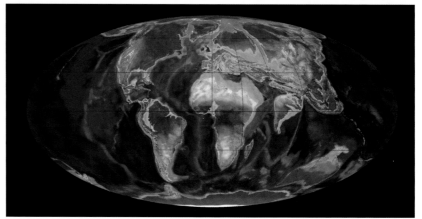

The long-term declining trend in sea level that characterizes the Caenozoic is due to a combination of changes in tectonic and in ocean circulation patterns that, over time, isolated Antarctica over the South Pole leading to the development of very low temperatures on that continent and a drop in average temperatures globally. This, in turn, resulted in the concomitant growth of northern and southern polar ice-caps, alpine glaciers worldwide, and continental glaciers that waxed and waned as a result of a complex of planetary factors, not all of which are well understood at present. Within the Early Caenozoic major falls in sea level occurred at the end of the Danian Stage (50 m or 165 ft) and again during the Early Oligocene (100 m or 330 ft).

Climatically the Palaeocene exhibited cooler and drier climates than the Late Cretaceous. Overall conditions were distinctly warmer and more equable than modern climates. Arid conditions, marked by evaporite deposits, characterized broad regions that extended at least to the 30° N and S latitudes. These framed a broad zone of warm, wet, tropical conditions around the Palaeocene equator within which extensive coal deposits were formed. North and south of this paraequatorial zone warm-temperate regimes extended almost to the poles where

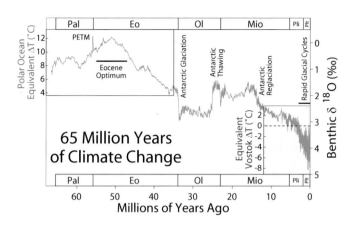

LEFT A summary of Caenozoic oxygen isotopic data with implications for global temperatures and various historical physiographic events indicated.

cool temperate conditions predominated. The geographic extent of Palaeocene warm conditions is signalled by the fact that crocodilians swam in the waters off Greenland and palm trees (with early primates in their fronds) occurred in what is now Wyoming. Reptiles – a biological proxy for warm conditions – were more widely distributed in the Palaeocene than they are at present. The Palaeocene was also characterized by very large reptiles including the champsosaurs and the so-called terror birds (large flightless birds). These were the top predators in many Palaeocene terrestrial environments. Tropical conditions extended much further north in Asia owing to the presence of a large shallow sea in that region – the remnant of the Mesozoic Tethys.

At the end of the Palaeocene the Earth's climate went through the warmest interval of the entire Caenozoic, the Palaeocene–Eocene Thermal Maximum (PETM, see above). Over an interval of *c.* 2,000–3,000 years in the Early Eocene, carbon isotopic data indicate that average global seawater temperatures rose by some 6°C (11°F). Various lines of evidence suggest that this temperature rise was associated with an increase in greenhouse gases in the atmosphere, especially carbon dioxide (CO_2) and methane (CH_4). Through the PETM the equatorial tropical belt expanded, pushing the subtropical arid zones to higher latitudes, as evidenced by northward and southward migration of Eocene evaporite deposits.

The remainder of the Middle and Late Eocene were characterized by falling global temperatures (with an average drop of about 4°C, 7°F) as evidenced, again, by foraminiferal isotopic data. During this time the Asian subtropical extension diminished as the Tethyan seaway shrank back in the face of falling sea levels. This interval ended in the Late Eocene with an abrupt and dramatic further fall in global temperatures of another 4°C or so, at or just after the Eocene–Oligocene (E–O) boundary, which steepened latitudinal temperature gradients sharply, especially in the southern hemisphere. This Early Oligocene temperature fall coincided with the sharp global sea-level fall (about 100 m, 330 ft) with all the accompanying ecological affects for marine and terrestrial organisms we have touched on above. It will come as no surprise to learn that both the PETM, and the Middle–Late Eocene cooling, were accompanied by species extinctions.

EXTINCTIONS

Curiously the only major organismal group affected by the PETM event was the benthic foraminifera, which lost between 35 and 50% of extant (Palaeocene) species within about 1,000 years. This observation has long been regarded as surprising insofar as benthic foraminifera were not affected strongly by the end-Cretaceous extinction event. Deep-water species were especially hard hit.

Other deep-dwelling marine groups (echinoids, bivalves, gastropods) all underwent diversifications during the PETM, as did all major marine plankton groups (nannoplankton, dinoflagellates, diatoms, planktonic foraminifera), many shallow-water marine groups (echinoids, bivalves, gastropods, sharks and rays, bony fish) and even many terrestrial groups (mammals, birds). Other than the benthic foraminiferal signal though, the most prominent effect of the PETM was to change the geographic ranges of many plant and animal species. Major northward and southward migrations took place as the planet warmed. Mangrove thickets and rain forests occurred as far north as Wyoming and Belgium and as far south as Tasmania. Ellesmere Island in the present-day Canadian Arctic sported turtles, hippopotami and palm trees, whereas members of the tropical dinoflagellate genus reached the North Pole. Temperate forests grew denser and wetter, a situation some mammal species adapted to by decreasing their size (e.g. *Hyracotherium sandrae*). Migrants from Laurasia included horses, rhinoceros, sheep and antelope, while a number of modern mammal species

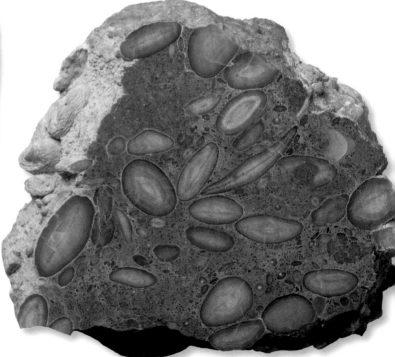

make their first appearance in the fossil record at or shortly after this event (e.g. elk, tapirs, rodents, bats, owls, elephants, whales).

The story is quite different, however, for extinctions in the Late Eocene and into the overlying Oligocene, the E–O extinctions. As with each of the other extinction events we have looked at in detail, the E–O extinctions took place over an extended time interval, in this case well over 10 million years in duration. As with earlier summaries, the Sepkoski dataset subdivides this interval into stages. Taking these data at face value the magnitude of the E–O extinctions is just a bit over 30%, much smaller than the extinction events described previously, but still a significant drop in overall generic richness. Using the reverse rarefaction method of Raup to translate this value into number of species, we can estimate a species-level extinction magnitude of approximately 60%. But owing to the fact that this is a comparatively young extinction we actually have a much more detailed understanding of the E–O extinction than this simple analysis would suggest. Prominent victims of this extinction event are shown here.

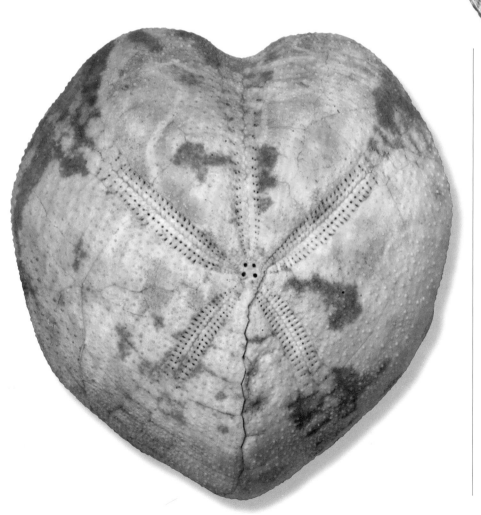

THIS PAGE Prominent victims of the Palaeogene extinction events: gastropods (above, *Athleta luctator* from rocks around Barton, England); echinoid echinoderms (left, *Micraster* sp.) and turtles (below, *Stylemys nebrascensis* from the western USA).

THIS PAGE Prominent victims of the Palaeogene extinction events: artiodactyl mammals (top right, *Andrewsarchus*, an unusual artiodactyl scavenger from Mongolia), archaeocete whales (right, *Protectus*, a primitive toothed whale from the Mediterranean), condylarth mammals (below, *Phenocodus*, a sheep-sized mammal with five digits on each hoofed foot, found in Europe and North America) and champsosaurs (bottom, *Champsosaurus gigas* from the Sentinel Butte Formation, western North Dakota, USA).

The extinction interval actually begins at the boundary between the Early and Middle Eocene (specifically the boundary between the Lutetian and Bartonian ages, about 40.4 million years ago). Over a relatively short time interval close to or spanning this boundary marine molluscs undergo a severe extinction event with some 89% of marine gastropods and 84% of marine bivalves disappearing from the fossil record. A few species of planktonic foraminifera also record their last appearances in this interval. Most deep-sea cores exhibit a hiatus at this boundary, which signals changes in the pattern of deep-ocean circulation, or the corrosiveness of deep-ocean waters with respect to carbonate dissolution, or both.

The next significant extinction interval is the Middle–Late Eocene boundary (the boundary between the Bartonian and Priabonian ages, about 37.2 million years ago). This interval represents a much more significant extinction event with almost 50% of marine coccolithophorid species vanishing from the fossil record. The majority of these were tropical species. Also, all tropical spinose planktonic foraminifera become extinct at or near this horizon. These extinctions suggest that relatively cooler water masses migrated into tropical areas where relatively warmer water masses had predominated formerly, an interpretation that agrees with isotopic data and the data from marine diatoms which do not exhibit an extinction event during this interval.

Higher up on the continental shelves the larger members of a diverse, formerly ubiquitous, group of benthic foraminifera – the nummulites – become completely extinct at this time, though smaller species of this group do struggle on to the end of the epoch. As a result, the disappearance of large nummulitid species represents an important time marker for the Middle–Late Eocene. Marine molluscs also succumb to another major extinction during the early Late Eocene with losses of some 70% of gastropod species and 60% of bivalve species. Finally, echinoid echinoderms – especially tropical forms such as those among the well-studied faunas of the north Gulf of Mexico/Caribbean – suffered significant losses in species numbers during the Middle–Late Eocene transition.

On the continents aquatic taxa such as crocodilians, champsosaurs and turtles all declined in numbers as the sea-level regression drained and reduced the area of their aquatic habitats. Terrestrial turtles (tortoises) increased in diversity as they were adapted for drier conditions. Archaic mammal groups such as taeniodonts, archaenodonts, uintatheres, nyctitheres, hyopsodont condylarths, anaptomorphine primates, sciuravid rodents, dichobunid artiodactyls, limnocyonid 'miacid', and mesonychid carnivores, mesonychids, and isectolophid tapiroids were also gone by the Middle–Late Eocene boundary. These were replaced in the Late Eocene by species of 'modern' aspect that were adapted to less forested and drier conditions, including early representatives of many familiar mammals (e.g. shrews, rabbits, gophers, squirrels, camels, dogs and rhinoceros, see Prothero 1994).

But this latest Eocene whimper after the Middle–Late Eocene debacle does not end the story of Middle Caenozoic extinctions. As I mentioned above, the main sea-level fall of the Caenozoic came in the middle part of the Early Oligocene. The reasons for this sea-level regression are described below. Here I want to confine myself to describing the effect it had on the floras and faunas extant at the time.

Beginning at the base of the marine food web, the taxonomic richnesses of tropical phytoplankton species crashed in the Early Oligocene. All but a single temperate-zone coccolith species that had appeared since the Early Eocene vanished, including several species that had evolved to fill vacant ecological roles in the wake of the Middle Eocene nanoplankton extinction event. Overall, this extinction eliminated *c.* 30% of the Middle Eocene species.

At the same time, cold-tolerant species increased in diversity. By the Late Oligocene the coccolith flora had adopted a new, highly provincialized structure based on differences in the characteristic temperature of water masses. This Early Oligocene evolutionary–ecological reorganization is also seen in diatoms and dinoflagellates, the former of which underwent a *c.* 45% Early Oligocene extinction. Among benthic foraminifera, the last of the small nummulitids succumbed in the Early Oligocene, precipitating the complete extinction of this major and formerly diverse group. Deep-sea textularid foraminifera – species that form their shells, or tests, from sediment particles they (sometimes selectively) picked up off the seabed and cemented in place – also underwent a significant extinction in the Early Oligocene, presumably due to changing deep-sea circulation patterns that reorganized the distribution of deep-sea water masses.

On the continental shelves the mollusc fauna was hit hard again, this time with over 95% of the gastropod species and 89% of the bivalve species that had survived or appeared since the Middle Eocene mollusc extinction vanishing from the fossil record by the end of the Early Oligocene. Similarly, echinoid echinoderms were decimated during this same time suffering over 65% extinction of extant Late Eocene species. Again, both the molluscan and echinoid extinctions affected tropical, warm-adapted forms severely whereas cold-tolerant species diversified in their wake. Perhaps the most spectacular victims of this extinction event, however, were the primitive archaeocete whales, which were replaced by larger, more specialized, odontocete (toothed) and mysticete (baleen) species that constitute the whale fauna of the present day.

TIMING

Although many scientific reports published in the 1970s and 1980s refer to the 'Terminal Eocene Event', improvements in the dating of Late Eocene sediments have shown that most of the groups previously referred to the Eocene–Oligocene extinction actually vanished from the fossil record at or around the Middle–Late Eocene boundary (Prothero and Berggren 1992, Prothero 1994). In the closing few hundred-thousand years of the Eocene the Earth was already populated by a well-diversified fauna with advanced body plans that were comfortable living in the savannah-like habitats that dominated the broad temperate regions. Crocodile, lizard and amphibian species continued their retreat to the equatorial tropics and new migrant species (especially from Asia) continued to appear in western Europe and North America. Across the E–O boundary we now know that extinctions were modest and confined to a few remnant tropical, warm-adapted groups that had managed to hang on through the Middle–Late Eocene transition (e.g. a few dinoflagellate species, and the striking, spinose planktonic foraminifera genus).

CAUSE(S)

THE PETM EVENT

The cause of the PETM remains somewhat mysterious at present. However, the isotopic data that define the event itself are key to understanding it (see right). The sudden shift to higher $\delta^{13}C$ isotopic ratios allows mass balance calculations for the atmosphere as a whole to be made. Based on these calculations, it is doubtful that input from any single source can account for the observed isotopic changes.

The most obvious potential source of the CO_2 whose injection resulted in planet-wide warming (for which there is abundant evidence) is volcanic eruption. There is a candidate large igneous province (LIP) of the correct age, a series of hydrothermal vents and intrusive igneous rocks thousands of kilometres in areal extent that exist in sedimentary basins along the margin of Norway and west of the Shetland Islands – the Brito-Arctic Volcanic Province. These eruptions are associated with tectonic rifting in the northern Atlantic Ocean. However, the amount of carbon expected to have been released naturally from eruptions on this scale is not sufficient to have caused the PETM. At least part of this difference might have been made up by the carbon-rich nature of the sediments beneath which these eruptions occurred. But even with this enhancement, the amount of carbon predicted to be released still falls short of the 1,500 gigatonnes that would be required (based on current mass balance models).

Debate continues regarding where the extra isotopically 'light' carbon that characterized the PETM event came from. One interesting suggestion is that emplacement of the Norway–Shetland volcanic complex raised marine water temperature enough to melt methane clathrates on the continental shelves. Clathrates are molecular 'ice cages' composed of one type of molecule that 'traps' another type of molecule in its structure (see p.152). For some time it has been known that large deposits of methane (CH_4) clathrates form naturally in the arctic permafrost and beneath the ocean bed.

Naturally occurring clathrates are only stable under a limited set of temperature and pressure conditions. Individual clathrates will thaw and release their trapped molecules whenever local conditions drift out of this stability field. However, in the case of methane clathrates the potential exists for a positive feedback mechanism to become established if global temperatures rise markedly and suddenly. Under

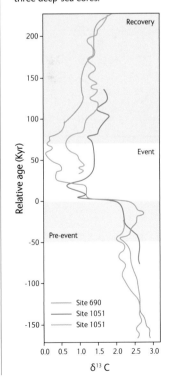

BELOW A summary of carbon isotopic data across the PETM from three deep-sea cores.

RIGHT Photograph of a methane clathrate on the deep-sea floor.

BELOW RIGHT A major source of methane gas comes from frozen hydrate crystals which exist on the sea floor in vast quantities. Due to very the cold temperatures and high pressures that have existed on the sea bed throughout the Holocene, these methane clathrates have remained frozen in place. However, if the climate changes the possibility is that these deposits may become unstable and melt, releasing large quantities of methane into the atmosphere where their effects are, at present, unpredictable.

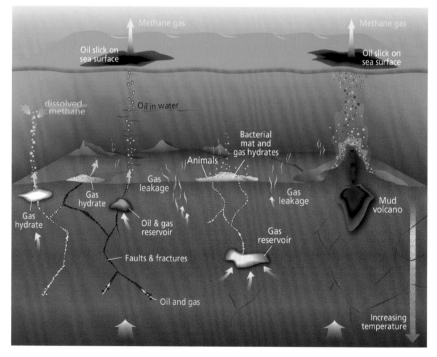

this scenario, a sudden temperature rise triggered by the volcanogenic emission of greenhouse gases could cause a large number of methane clathrates to melt, resulting in the injection of a vast quantity of methane (a potent greenhouse gas) directly into the atmosphere. This would result in further temperature increase, which would result in the melting of additional clathrates, and so on until the reservoir of available clathrates was exhausted or some other limiting mechanisms started up (e.g. extraction of greenhouse gases from the atmosphere due to enhanced rates of photosynthesis, most likely by marine phytoplankton).

Since methane clathrates are composed of isotopically light carbon this is seen by some researchers as the most plausible mechanism to account for the PETM event. Certainly raw mass-balance calculations and crude simulations that assume the availability of a large volume of methane clathrates suggest this mechanism could work in principle (see Zachos *et al.* 2005). Nonetheless, since clathrate melting leaves no direct stratigraphic evidence it is not possible at present to know whether the assumptions on which these calculations and simulations were based are correct. Additionally, there is the matter of why this mechanism has not operated at other times during Earth history. For the time being the 'clathrate gun' hypothesis should probably be regarded as an interesting suggestion, but one that must await the development of some way of testing the operation of this putative mechanism specifically before it can be either accepted or rejected as a potential cause of the PETM (or any other) light carbon isotopic excursion.

THE EOCENE-OLIGOCENE EVENT

Although the cause(s) of the PETM remain debatable, the causes of the Eocene–Oligocene extinctions are much better established. Following the PETM the Earth's climate entered the warmest phase of the entire Caenozoic and one of the warmest intervals in all of post-Proterozoic Earth history. This warm interval culminated in the Early Eocene, just before which another warm perturbation – the Eocene Thermal Maximum 2 (ETM-2, *c.* 53.7 million years ago) occurred. This second event was proximally caused by the injection of isotopically light carbon into the atmosphere and is thought to result from the same class of mechanism(s) that caused the PETM event. Following the Early Eocene peak warming (*c.* 50.0 million years ago) the planet's climate began a protracted cooling. Once peak warming had been reached extinctions begin to occur during times of instability in the cooling trend, first around the time of the Early–Middle Eocene transition (48.6 million years ago) and later at the Middle–Late Eocene transition (*c.* 40.4 million years ago). The Eocene–Oligocene boundary coincides with a period of temperature instability, but one that followed from at least two previous disruptions to the overall cooling trend. Then, in the Early Oligocene, a major fall in oxygen isotopic values signals what can only be interpreted as the formation of a deep and long-lasting ice-cap over the southern pole. But what factors ultimately caused the temperature instabilities and ice-cap formation that caused global extinctions? Two models have been proposed, each tied ultimately to plate tectonics and volcanism.

At present the concentration of greenhouse gases in the atmosphere depends on interactions between factors that release these gases into the atmosphere and factors that store them in natural repositories. Ignoring exotic inputs such as those resulting from clathrates and/or human technology, volcanism is the primary release mechanism and storage in sediments, soils and/or plant tissues/skeletons the primary sequestration mechanism. Over the course of the Middle and Late Eocene palaeomagnetic data indicate that rates of sea-floor spreading declined following rearrangements of spreading directions. A decline in the production of new oceanic crust would presumably be accompanied by a decrease in rates of volcanism at the mid-ocean spreading centres and so a decrease in the input of greenhouse gases into the atmosphere. All other factors being equal, this reduction would be expected to result in development of a long-term cooling trend.

The Himalayan Connection

However, all other factors were not equal over the course of the Middle Eocene–Early Oligocene interval. Several other major tectonic events were taking place. For one, the Indian plate was colliding with the Asian plate and so forcing the Himalayan mountain range to rise out of the shallow seaway between the two continents. Episodes of mountain building increase the rate of physical and chemical weathering which, in turn, draws CO_2 out of the atmosphere as this compound combines with newly exposed chemically reactive minerals. In this process the CO_2 taken from the atmosphere is ultimately stored in soils along with CO_2 removed as a by-product of photosynthesis. In the oceans atmospheric CO_2 is stored in carbonate rocks formed from the remains a billions of microfossil skeletons. The rise of Himalayan and other mountain chains would remove CO_2 from the planet's atmosphere at a time when less CO_2 was being pumped in by volcanic processes. Once again, a long-term cooling trend would be the expected result.

Isolation of Antarctica

Still, there is a third factor to be considered – Antarctica. In the Early Eocene the southern tip of South America and southern coast of Australia were still joined, or very close, to Antarctica (see left). With the southern continents in this configuration warm surface currents from the tropical Atlantic and tropical Pacific would be channelled along both sides of Antarctica, transporting their load of tropical heat southward. These warm surface currents moderated Antarctic temperatures and thus those of the entire globe. However, as both South America and Australian moved north over the course of the Eocene the Tasmanian Seaway and Drake Passage became broader and deeper, so much so that a circum-Antarctic gyre began to form. Over time this band of water, perpetually circling the southernmost continent, became less susceptible to heat transfer from warmer currents

BELOW Palaeogeography of the Oligocene (c. 30 million years ago) world showing positions of continental landmasses and ocean basins.

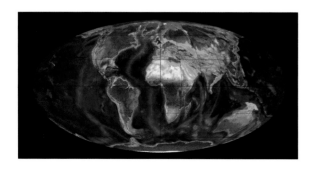

flowing south from the lower latitudes. As a result, conditions in Antarctica began to grow cooler. These cold polar conditions cooled the atmosphere and eventually the oceans, thereby refrigerating the entire planet. Based on the current oxygen isotopic record, the onset of this long-term cooling trend appears to have taken place in the Middle Eocene.

As the drift of South America and Australia away from Antarctica proceeded, the gyral organization grew stronger and refrigeration of the continent grew more intense, causing continental glaciers to form, sea level to fall, and the planetary albedo to increase; all factors that have been associated historically with both marine and terrestrial extinctions. This effect grew even stronger as deep-ocean passages grew deeper, first between Australia and Antarctica and later between South America and Antarctica.

Not only this, but as the surface waters offshore of Antarctica grew cooler than the deeper waters they began to sink, causing reorganization of vertical as well as horizontal marine water-mass circulation. This had two noteworthy affects. First, warmer nutrient-rich waters began coming to the surface offshore of Antarctica supplying the organisms that lived there with more food resources. Phytoplankton blooms ensued and, over time, a diverse ecosystem that included everything from protist-grade microalgae to new whale species developed to take advantage of the new nutrient source. Competition from these new species inevitably put pressure on more archaic species leading to their eventual extinction (e.g. archaeocete whales).

Second, coupled with tectonic events in the North Atlantic (e.g. sinking of the Iceland–Faroe Ridge) a complete system of vertical circulation was established throughout the ocean basins – the marine conveyor (see p.58). This aerated the waters of the ocean basins, changed the physical characteristics of these waters, opened new areas of the ocean floor for colonization by organisms and, also inevitably, put pressure on deep-sea species that had adapted to the previously stable suite of conditions. All of these effects would be expected to result in widespread extinctions and concomitant replacement of ecological roles with newly appearing species.

Non-tectonic volcanism was also part of the Late Oligocene story in the form of emplacement of the Ethiopian–Yemen Volcanic Province (see p.156). This eruption was centred on the Afar mantle plume and emplaced some 2 million km^2 (1.3 million mi^2) in present-day Ethiopia, Yemen, Djibouti, Saudi Arabia, Sudan and Egypt over a 1 million year interval at or close to 29.5 million years ago. This is the approximate age of the Early–Late Oligocene boundary and post-dates the refrigeration of Antarctica along with the Early Oligocene extinctions. The Ethiopian–Yemen eruptions coincide approximately with an elevation of the benthic foraminifer-derived oxygen isotopic ratio that, in turn, signalled onset of warmer global temperatures throughout the Late Oligocene. The somewhat curious failure of such massive volcanic eruptions in northeastern Africa to elicit a stronger extinction response may be the result of the biota present at that time already being composed of extinction-resistant species that had just survived a major global environmental perturbation.

RIGHT Map of the Ethiopian-Yemen Volcanic Province (lava fields shown in red).

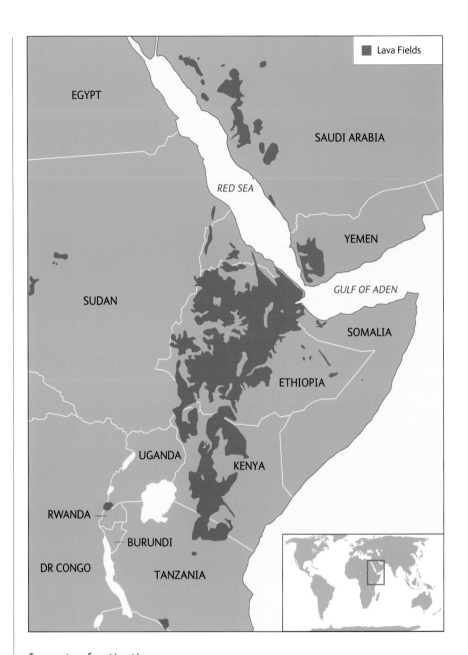

| Lava Fields |

EGYPT

SAUDI ARABIA

RED SEA

YEMEN

GULF OF ADEN

SUDAN

SOMALIA

ETHIOPIA

UGANDA

KENYA

RWANDA

BURUNDI

DR CONGO

TANZANIA

Impacts of extinctions

Before ending the discussion of potential causes of the Eocene–Oligocene extinctions mention needs to be made of the idea that a bolide impact at the E–O boundary may have been responsible for some or all of the extinctions observed in this interval. In the wake of the original 1980 article announcing the bolide impact scenario for the Maastrichtian extinction by the Alvarez team rumours began to circulate of other iridium anomalies that had been found at other geological boundaries, including the E–O boundary. This E–O anomaly was later announced formally by the same team that discovered the K–Pg iridium anomaly. However, the Eocene anomaly did not occur coincident with the E–O boundary, but in the middle of the Late Eocene.

More to the point, it did not coincide with any stratigraphic horizon that could be linked to a large number of extinctions. The Late Eocene iridium anomaly was also much smaller than the one that had been recovered from the Italian K–Pg boundary section.

Later, the crater associated with this anomaly and other Late Eocene impact debris was discovered by a team of earth scientists investigating the geological structure of subsurface sediments in the Chesapeake Bay area on the eastern seaboard of the USA. The Chesapeake Bay crater is estimated to be 100 km (60 mi) wide and 35.5 million years old. This is a significant impact feature that must have devastated the local region at the time of its formation. This crater, plus a small (15-km or 9-mile diameter) crater of the same age that was discovered on the continental shelf off the state of New Jersey, scattered debris over an area that stretched from the Caribbean as far south as the South Atlantic and as far east as the Indian Ocean. In 1975 a third Late Eocene crater was discovered in Siberia. This is the Popagai crater, which is 90 km (56 mi.) wide and has been dated at 35.7 million years ago.

These three impact craters constitute direct evidence for cataclysmic perturbations of the biosphere at least regionally. Given the size of the Chesapeake and Popagai craters, and the analogies drawn with other extinctions under David Raup's 'kill-curve' model (see right), a significant reduction in the standing crop of terrestrial and marine species should have been expected. But, as we have seen, this prediction has not been confirmed by studies of the fossil record. The only extinctions found associated with the Late Eocene iridium anomaly were those of five radiolarian species recovered from cores taken in Caribbean sediments. None of the extinctions now known to have occurred during the Early–Middle Eocene transition, or the Middle–Late Eocene transition, and certainly none of the extinctions linked to the Early Oligocene temperature and sea-level drops, can be causally linked to any of these E-O impacts. More tellingly, the fact that such large impacts (apparently) can occur and result in such minor levels of extinction (for me) casts doubt on the entire prospect of bolide impact as a simple extinction mechanism.

Of course, small volcanic eruptions can and do occur and (apparently) do not precipitate larger extinction events. But the Chesapeake and Popagai craters are not small impacts in terms of size - Popagai is the fourth largest crater known and Chesapeake one of the top 15 largest craters. If their diameters are added together the combination is only slightly smaller than the Chicxulub crater. Moreover, these impacts occurred in the same stratigraphic stage, the Priabonian which is the final stage of the Eocene Series. Still, in terms of extinction intensity the Late Eocene ranks 58th out of 77 stage-level subdivisions in the Sepkoski database. Based on these (and other) data from the Late Eocene (e.g. see MacLeod 2004,2005) I believe the assumption that the occurrence of a large bolide impact necessarily caused widespread extinctions simply does not hold up in the face of scientific scrutiny of evidence provided by the fossil record.

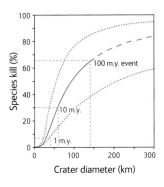

ABOVE Kill curve of David Raup relating crater size to percent of species killed. (Redrawn from Raup 1992.)

13 The Neogene and Quaternary extinctions

THE FIRST TWO EPOCHS OF THE NEOGENE SYSTEM, the Miocene and Pliocene, were decidedly quiet times in extinction activity, but far from lacking in terms of both global and major regional events. Biotically the Miocene and Pliocene were times of radiation for the lineages that contain most of the species we encounter in the modern world. Slightly more primitive versions of such familiar groups as gastropods, bivalves, sharks, rays, fish, birds, antelope, rhinoceros, horses, cats, elephants and primates are all recognizably present in the Neogene fossil record (see below) as well as more exotic extinct species such as the giant shark and the giant perissodactyl chalicotheres (horse and rhinoceros relatives).

OPPOSITE Red Rock Canyon State Park, California, USA. These Neogene sediments contain terrestrial fossils and are located 97 km (60 mi.) from the present coast, suggesting that sea level stood tens of metres higher earlier in the Neogene than it does today.

Reconstruction of Miocene marine and terrestrial environments.

LEFT A Miocene deep-ocean environment including an ancestral great white shark, *Carcharodon megalodon*, attacking a juvenile whale. Fossil evidence of its teeth and a few bones suggests that *C. megalodon* grew over 20 m (66 ft) in length and weighed over 100 tons. It is the largest carnivore that ever lived.

BELOW LEFT A Miocene terrestrial environment from Chad. Foreground: the hyaena *Ictitherium* (left), the sabre-toothed cat *Machairrodus* (centre), the giraffid *Sivatherium* (right). Background: the early hominid genus *Sahelanthropus* (left), the perissodactyl *Hipparion* (centre), the tetraconodontid pig *Nyanzachoerus* (centre), the antelope *Kobus* (centre) and a band of cercopithecoid monkeys (upper right).

THE MIOCENE

SETTING

Geographically the continents had assumed positions very close to their present configuration (see below). The land bridge between North America and South America was not present in the Miocene, but formed in the Pliocene. India continued its collision with Asia pushing up the Himalayan mountain range and lifting the whole of the Tibetan Plateau half a kilometre into the sky. Africa also continued its collision with Eurasia pushing up mountain ranges around the Mediterranean Sea which, along with the Black Sea, was the last remnant of the formerly vast Tethys Sea. These changes aside, looking at the Earth from outer space at this time the most evident difference would be the waxing and waning of ice-caps at the North and South Poles. On the whole sea level stood low throughout the Neogene, but exhibited pronounced fluctuations as the extent of the polar ice-caps, the Greenland ice-cap and alpine glaciers grew or retreated in response to global temperature fluctuations (see p.145).

Climatically, the Early and Middle Miocene was a time of relatively warm average global temperatures. From the Middle Miocene on to the present day though, the Earth entered another cold phase, as indicated by oxygen isotopic data obtained from deep-sea benthic foraminifera (see p.145). These isotopic ratios record even colder conditions in the Pliocene than those recorded during the initial refrigeration of Antarctica in the Early Oligocene. Owing to cooling of the poles latitudinal temperature gradients were as steep as they are today. Ecologically the Earth's flora and fauna were subdivided into reasonably distinct provinces by temperature, patterns of seasonality, precipitation and barriers to migration.

Particularly noteworthy physical events that occurred during this interval of Earth history include the following.

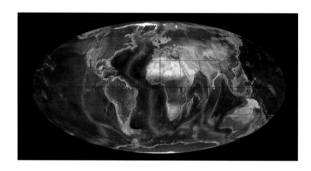

ABOVE Palaeogeography of the Miocene (*c.* 20 million years old) world showing positions of continental landmasses and ocean basins.

- **Expansion of the Antarctic ice-cap.** The Earth first acquired an extensive Caenozoic south polar ice-cap in the Late Eocene and this has remained a constant feature of the planet for some 34 million years. Physical and biological evidence indicates the ice-cap formed in two stages. The initial stage of formation coincided with the tectonic isolation of Antarctica (see p.154) and resulted in growth for the continent's Western Ice Sheet. Oxygen isotopic data from Antarctic benthic foraminifera indicates that, during the Middle Miocene, a second drop in average ocean temperatures occurred. This interpretation is also supported by an increase in the width of the belt of sediment dominated by siliceous microfossils surrounding that island continent. Mapping the extent of the zone of silica-rich sediments around the South (and the North) Pole provides a good proxy indication of the regional extent of cold conditions. The final line of evidence for Middle Miocene cooling is the geographic extent of ice-rafted debris around Antarctica. All three of these proxies indicate that the zone of cold water around Antarctica increased during the Middle Miocene. This cooling resulted in growth of the East Antarctic Ice Sheet. Curiously, temperature indicators in low latitudes during this time indicate stable, warm conditions prevailed. Thus, latitudinal temperature differences must have been greater in the Middle Miocene than during preceding Caenozoic time intervals. A second round of Miocene Antarctic cooling appears to have taken place in the Late Miocene (5–6 million years ago) based, again, on oxygen isotope data, the extent of silica-rich sediments surrounding Antarctica and, again, the extent of ice-rafted debris.

- **Disappearance of the Mediterranean (=Tethys) Sea.** As Antarctica was progressively buried under a continent-sized ice sheet sea level decreased globally. This had an especially profound effect on the Mediterranean Sea insofar as a short-term sea-level drop in the terminal stage of the Miocene Series (the Messinian) that took the level of the Atlantic Ocean below the depth of the sea floor at the Strait of Gibraltar. So long as water could flow from the Atlantic Ocean into the Mediterranean

Basin to replace Mediterranean water that evaporated because of the hot, dry conditions there, the level of the Mediterranean could be maintained. Exposure of the Camarinal Sill in the Strait of Gibraltar effectively cut off the flow of water into the Mediterranean Basin. The result of this isolation was that the level of the Mediterranean Sea began to fall (right). Over the course of the Messinian Epoch the entire sea disappeared completely as evidenced by geological cores obtained by deep-ocean drilling ships that have recovered Miocene salt deposits from the middle of the Mediterranean Basin. In addition, subsurface mapping has located deep river-cut valleys below the present tracks of the Rhone, Po and Nile rivers, all of which flow into the Mediterranean. Indeed, the depth and extent of the Nile river valley during the Late Miocene is estimated to rival that of the Grand Canyon in the western US. Rising sea levels in the Early Pliocene caused by glacial melting breached the Strait of Gibraltar, refilling the Mediterranean Basin with seawater and re-establishing communities of marine organisms there.

ABOVE Artist's conceptualization of the Messinian Salinity Crisis during which the Mediterranean Sea became isolated from the Atlantic Ocean and dried up.

- **Emplacement of the Columbia River Volcanic Province.** This event was centred around the present-day Yellowstone mantle plume and constitutes a lava field track formed as the North American plate passed over the plume (see p.162). The province consists of about 175,000 km³ (110,000 mi³) of basaltic lava that was erupted during an interval from 16.5 to 14.5 million years ago with smaller eruptions continuing to 6.0 million years ago. This volcanic event coincides in a general manner with a slight extinction-intensity peak recorded in the Sepkoski dataset for the Middle Miocene.

RIGHT Columbia River Volcanic Province, Palouse Canyon, southeastern Washington State, USA showing the scale of the basaltic lava flows.

BELOW Map of the extent of the Columbia River Volcanic Province laval fields (shown in red).

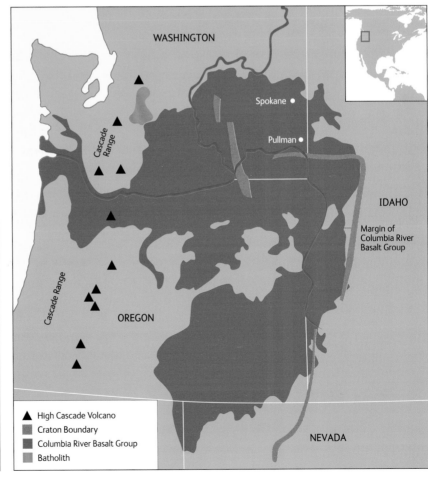

WASHINGTON

Spokane ●

Pullman ●

Cascade Range

Cascade Range

IDAHO

Margin of Columbia River Basalt Group

OREGON

NEVADA

▲ High Cascade Volcano
  Craton Boundary
  Columbia River Basalt Group
  Batholith

- **Spread of grasslands.** Grasses originated in the Late Cretaceous, but were a minor component of terrestrial floras in most biomes until the cooling and drying of the climate that took place in the Late Oligocene. Grasslands, prairies and savannahs were well established by the Middle Miocene. Some of the extinctions recorded within this interval may have been the result of competitive elimination of early forms of modern mammals (including horses, deer, camels, rhinoceros, antelopes) by species with advanced dental and physiological designs that were better able to cope with the nutrient-poor and abrasive forage offered by the vast Miocene grass fields.

EXTINCTIONS

Raup and Sepkoski (1984, 1986) advanced the idea that a significant extinction event took place in the Middle Miocene. The Sepkoski genus-level dataset suggests that, lumping all extinctions occurring in all Miocene stages together the overall level of extinction assigned to this time interval is only 21% of then extant genera, far lower than any extinction event we have considered thus far. This level of generic loss implies a species extinction magnitude of *c.* 40%. Of course, because the Sepkoski dataset lumps some 17 million years into a single time interval the extinction rate observed at any one time during this interval would be well below 21% of genera and 40% of species. In the oceans a single investigator has reported an extinction of between 38 and 52% of planktonic foraminifera species during the early Middle Miocene, but this questionable event apparently involved no other marine animal groups. An early Late Miocene extinction event has also been proposed for terrestrial mammals extending into the late Late Miocene which involved relatively large ungulates of various sorts, both selective browsers and non-selective grazers (e.g. peccaries, horses, rhinoceros).

The lesson that can be drawn from the Miocene extinction record is that this event is similar to that of the Late Oligocene. In the Miocene we see evidence for profound changes to the Earth's climate, marine circulation patterns, sea level and physiography at local, regional and global scales, all of which occurred without precipitating a major extinction event. Extinctions did take place throughout this interval. But the overriding story is one of radiation and diversification in most organismal lineages rather than contraction and extinction.

THE PLIOCENE

The Pliocene Epoch represents the prelude to the Recent, the time interval during which the ecological stage, the species and the roles of the organisms who populate our modern world began to assemble. The marine fauna of the Pliocene was modern in aspect and contained many extant species. This is in keeping with the name Pliocene, which means 'continuation of the Recent', though Charles Lyell was referring specifically to the molluscan fauna when he named the interval.

SETTING

As a result of the availability of Pliocene mammalian fossils and because of their diversity, more attention has been paid to this group than to any other in this part of the fossil record, even to the extent that contemporary palaeontologists know the Pliocene mammalian fauna that existed on different continents in a fair amount of detail. In North America weasels, opossums, dogs, bears and proboscideans all diversified while ungulates declined and sloths, armadillos and glyptodonts migrated in from South America. Once the land bridge between North and South America had been established the Great American Faunal Interchange began, which meant that a few South American species migrated north and did well (e.g. sloths, armadillos, glyptodonts, coatis) while a large number of North American species invaded the south and, with the exception of a few obscure groups (e.g. macrauchenids, toxodonts), drove much of the indigenous mammalian fauna

Reconstructions of Neogene terrestrial and marine environments.

ABOVE RIGHT Mammals of the Miocene Era based on deposits recovered from the Iberian Peninsula. Foreground: a prehistoric pig, *Bunolistriodon lockarti*, lower left) with its young, and a family group of *Gomphotherium angustidens* (upper right and background), wading through shallow water.

BELOW RIGHT Pliocene marine environment based on fossils from North Carolina, USA. Foreground: a school of the bonito, *Sarda sarda*, small fish, and a Caribbean monk seal, *Monachus tropicalis*, lower right. Background: the short-fin mako shark, *Isurus oxyrinchus*, centre.

to extinction there. Unlike South America, Australia remained an isolated island continent. Accordingly, marsupials continued to dominate the Australian mammal fauna. The only significant placental migrants to become established in Australia were Asian rodent species though a few species indigenous to Australia also appeared at this time (e.g. platypus).

Europe saw the appearance of many immigrant species including primates, hyraxes, hyaenas and sabre-toothed cats from Africa. Asia bore witness to the introduction of camels and proboscideans from North America. Africa was the home of many ungulate species, which continued to diversify there, partly in response to the introduction of many species from North America, including many species regarded today as being 'characteristically' African (e.g. camels, giraffe, horses, rhinoceros). Africa also saw the evolution of the first hominins at *c.* 3.0 million years ago.

With regard to other terrestrial or semi-aquatic vertebrates, while the large, flightless (phorusrhacid) birds of South America declined in the face of competition from North American mammals, one species, *Titanis walleri*, migrated into North America. Judging from the number of its fossils that occur in Late Pliocene sediments *Titanis* was quite successful there. Also, crocodiles disappeared from Europe as the climate cooled, but snakes diversified on all continents, most likely as a result of the diversification of small mammals. Dominant among these radiating small mammal groups were rodents, which underwent an evolutionary radiation on all continents as well as being frequent migrants. Grasslands continue to expand during the Pliocene in response to the cool, dry climate, at the expense of temperate deciduous forests.

Geographically the continents had assumed positions very close to those they occupy today. The Mediterranean Basin was dry at the beginning of the Pliocene, but refilled with marine waters from the Atlantic during the Early Pliocene sea-level rise. As mentioned above, the Isthmus of Panama emerged from the Caribbean Sea at *c.* 3.5 million years ago facilitating the Great American Faunal Interchange, the height of which took place during the Late Pliocene. The Late Pliocene was also a time of significant volcanic activity with the emplacement of two LIPs in western and central Europe and in northern and central Africa. The precise timing and extent of these LIP events have not been established, but both are regarded as having taken place in the Late Pliocene–Early Pleistocene.

Sea level stood slightly higher (*c.* 25 m) on average during the Pliocene relative to today, but fluctuations in sea level were frequent and pronounced, increasing in amplitude towards the end of the interval. Overall, the Pliocene climate was cooler and drier than that of the Miocene and more similar to the climate of the present day. The global cooling that accompanied refrigeration of Antarctica most likely boosted by other factors (e.g. increased albedo, changes in marine circulation patterns), resulted in the formation of a north polar ice-sheet. This was intermittent at first, but eventually became a continuous feature of the planet, an event that was also associated with development of the first permanent ice-sheet in Greenland at *c.* 3.0 million years ago.

EXTINCTIONS

As with the Miocene, the number of Pliocene extinctions recorded by the fossil record is quite low despite what appear to be wrenching physical and climatic changes in the Earth's environment. The Sepkoski database indicates that global extinctions over the entire Pliocene interval were only 8.5% of the standing fauna, markedly less even than those of the Miocene. As indicated by the discussion of vertebrate faunas (above), the extinctions that did occur were largely local, resulting from the disappearance of species on a single continent or a local region, but not globally. Numbers of extinctions were more severe among marine invertebrates than among terrestrial vertebrates. Overall the Pliocene was a time of higher species richness than is present in the modern world. But this pattern of decline is strongly regionalized. According to the noted palaeontologist Stephen Stanley, 80% of Early Pliocene bivalve species occurring in the Atlantic and Caribbean are extinct today, though this percentage drops to 37% for the Pacific faunas of California and Japan. Comparing these two biomes Stanley estimated that western Atlantic bivalves and gastropods suffered an excess extinction intensity of some 66% over the Pliocene molluscan background rate.

Due to the large number of fossiliferous sections and cores available for study by palaeontologists, very detailed regional data are available for very recent time intervals. Together these records make an important point about more ancient events. In all likelihood strong regional differences were part of the context of these older events just as the documented record indicates they are for younger events. However, owing to limitations in access to sections and cores for study, as well as lack of familiarity with both the taxonomy and ecologies of species, genera and, for the oldest events, families, orders and classes, the regional signals for the older events have either not been preserved or have gone largely unrecognized to date. We cannot resurrect these ancient regional signals. But we can filter the Neogene extinction data and ask ourselves 'What would these extinctions look like if the record available for study was similar to that of the Palaeozoic and Mesozoic extinction events?'. When these simulations are run the magnitude of the Miocene and Pliocene extinction events drops to the level where they are barely recognizable.

THE PLEISTOCENE

The youngest geological series of Earth history is the Pleistocene which lasted from 2.59 million years ago to a scant 11,700 years ago. This time period is best (but erroneously) known as the 'Ice Age' as it is characterized by very cold temperatures and extensive continental glaciation. The Pleistocene is also the interval in which an extinction event – the Pleistocene Extinction, also referred to as the Pleistocene Megafaunal Extinction – occurred. This extinction has been referred to as a 'mass extinction' though, as we will see, it is actually a very small extinction event compared to the 'Big Five' from more ancient times in the fossil record.

SETTING

The Ice Age label for the Pleistocene is incorrect for two reasons. First, the Pleistocene interval contains sub-intervals of extreme glacial conditions and sub-intervals of milder conditions, the interglacials. Glacial intervals are characterized by cooler average air and sea temperatures, lower average rainfall, more extensive sea ice, advances in both continental and mountain glaciers, expansion of permafrost zones and lowered sea levels. Interglacial intervals are characterized by warmer average air and sea temperatures, higher average rainfall, less extensive sea ice, retreats in both continental and mountain glaciers, reduction in permafrost zones and raised sea levels. In extreme cases the difference between glacial and interglacial intervals has been marked by instances of continental glaciation as far north and south as 40° latitude and sea-level changes of over 100 m (330 ft). At present the Earth is in an interglacial interval and has been for the last 11,000 years. Second, the onset of extreme glacial conditions occurred in the Late Pliocene, not the Pleistocene . Of course, extensive continental ice sheets have also characterized other intervals of Earth history as well, notably the Precambrian, Late Ordovician/ Early Silurian, Late Devonian, and Carboniferous-Middle Permian.

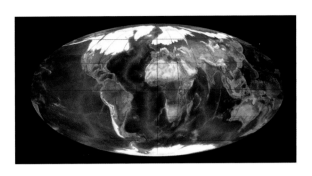

ABOVE Palaeogeography of the Pleistocene (*c.* 50,000 years before present) world showing positions of continental landmasses and ocean basins.

The transition between a largely ice-free world and one in which as much as 30% of the planet's surface was covered with ice (see above) in a periodic series of glacial advances and retreats was a relatively sudden, but not a discontinuous process. The onset of Antarctic glaciation began as far back as the Early Oligocene (*c.* 33 million years ago) (see p.161). Throughout this entire interval the Earth's climate oscillated between relatively cooler and relatively warmer phases. Following the Middle Miocene expansion of the southern polar ice-sheet, however, this short-term oscillatory climate pattern was superimposed on a steep, long-term trend towards colder average global temperatures.

Beginning around 3.0 million years ago the character of Earth's climate changed in two ways. First, the long-term trend of decreasing global average temperatures grew steeper resulting in refrigeration of both the polar and middle latitudes with an associated steepening of the latitudinal gradient in temperature differences. Second, the amplitude of the short-term global temperature oscillation increased resulting in the development of glacial intervals that were much colder than in previous times and interglacial intervals that were much warmer. Through the Early Pleistocene the period of these glacial–interglacial cycles was *c.* 40,000 years, but about 1 million years ago this period lengthened to *c.* 100,000 years (see p.168).

Why the Earth entered this clock-like state of cycling is not understood completely. It is likely that the short-term oscillations between relatively warm and relatively cool conditions has always been part of the Earth's climate system, but are often obscured in older sediments owing to poor preservation, lack of access to long, continuous sequences of sediments and lack of method to date sediments in

RIGHT Detail of Pliocene and Pleistocene climate change data derived from benthic foraminiferal oxygen isotopic data.

BELOW RIGHT Milankovitch cyclicity modes. Axial tilt (blue), orbital eccentricity (green), longitude of perihelion (magenta), orbital precession (red), isolation (grey). Two empirical climate proxy records that reflect the combination of Milankovitch cyclicities are shown at the bottom of the chart.

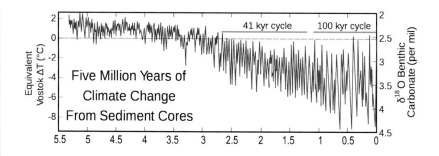

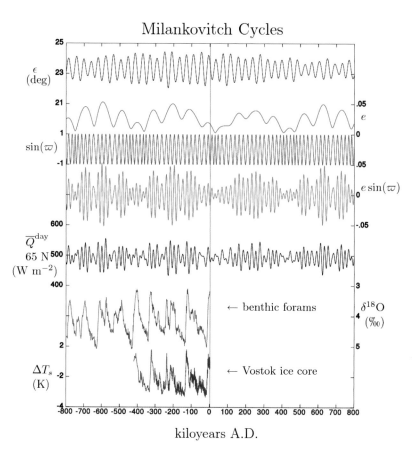

different regions with a precision that enables the cycles to be recognized. However, it has been known since the 1920s that variations in the Earth's orbit around the sun – variations in the orbit's eccentricity, the tilt of the Earth's rotational axes and the geometric relation between these orbital parameters – result in periodic variations in the amount of sunlight reaching all parts of the Earth's surface. These variations, whose periods are 21,000 years (precession of seasons), 26,000 years (precession of Earth's rotational axes) and 41,000 years (oscillation of axial tilt) would be expected to affect the Earth's climate, by inducing cyclic variation in average temperatures, rates of erosion, atmospheric composition, weather patterns, patterns of marine and atmospheric circulation, etc., all of which affect climate.

Evidence indicating the operation of this Milankovitch cyclicity comes from many sources throughout the Neogene and Palaeogene fossil and stratigraphic records (see left). Moreover, as these data and geochronologic resolution improves for older intervals of Earth history, Milankovitch-like patterns of variation are being recovered from earlier times. The longer-term trends seen in the Neogene record probably have their origin in various feedback mechanisms that either enhance or inhibit the establishment of temperature trends (e.g. albedo, sea level, volcanism-induced changes in the composition of the atmosphere, the positions of continents, marine circulation patterns, occurrence and orientation of mountain ranges). Because of the obvious desire to improve our ability to predict changes in the Earth's climate, understanding the processes that have historically controlled the planet's climate is a very active area of earth science research and will likely remain so for the foreseeable future.

EXTINCTIONS

What effect did these changes have on Pleistocene species? The answer, put simply, is, 'Not much!'. Average rates of extinction for the Pleistocene interval as a whole are less than 2%, one of the very lowest extinction levels recorded for any time interval in Earth history. Species extinctions did occur in the Pleistocene. But like the Pliocene, the Pleistocene fossil record is dominated by regional disappearances of local populations — extirpations — not global extinctions of entire species.

Owing to controversy regarding the cause of Pleistocene extinctions in North America (see below) many believe this extinction event only affected that continent. This is incorrect. Extinctions did occur across the globe. While controversy surrounds most questions related to the Pleistocene regional extinction events, the single aspect on which everyone seems to agree is that it differentially affected large-bodied species such as horses (*Equus* spp.), the saiga antelope (*Saiga tatarica*), woolly mammoth (*Mammuthus primigenius*), the American mastodon (*Mammut americanum*), the American lion (*Panthera atrox*), a cave lion (*Panthera spelaea*), sabre-toothed cats (*Homotherium serum*, *Smilodon fatalis*), a short-faced bear (*Arctodus simus*), a glyptothere (*Glyptotherium floridanum*), ground sloths (*Glossotherium harlani*, *Megalonyx jeffersoni* and *Paramylodon*), peccaries (*Platygonus compressus*, *Mylohyus nasutus*), the Western camel (*Camelops hesternus*), a llama (*Paleolama mirifica*) and moose (*Cervalces scotti*). Overall, North America suffered the true extinction of 70% of the extant megafauna (mammals with a body weight of more than 5 kg).

In South America losses included a giant armadillo (*Holmesina*), glyptodonts (*Glyptodon*), mastodons (*Haplomastodon*, *Cuvieronius*), several ground sloths (*Mylodon, Glossotherium, Nothrotherium, Megatherium, Eremotherium* and *Catonyx*), a notoungulate (*Toxodon*), horses (*Hippidion*, *Equus*) and a sabretooth cat (*Smilodon*). Overall South America lost 80% of its extant megafauna. In Australasia losses included giant wombats (*Diprotodon*, *Phascolonus gigas*), marsupial tapirs (*Palorchestes azael*, *Zygomaturus trilobus*), short-faced kangaroos (*Sthenurus*, *Simosthenuru*, *Procoptodon goliah*), a large wallaby-like kangaroo (*Protemnodon* sp.), the marsupial lion (*Thylacoleo carnifex*) and a giant echidna (*Zaglossus hacketti*). Overall Australia lost 81% of its extant megafauna. Extinctions also occurred on many islands at this time including Cyprus, Crete, Majorca, Minorca, Hawaii, Madagascar and New Zealand.

Only a few true extinctions were recorded for the megafauna from Europe, Asia and Africa, but many local species populations vanished from these areas. In addition to these mammals, 19 genera of North American birds disappeared from the fossil record. A not inconsiderable number of smaller terrestrial vertebrates also suffered both global and regional extinctions though the numbers involved here do not add up to impressive

BELOW Prominent victims of the Neogene extinctions: marsupial lion (below left, *Thylacoleo carnifex*); giant beaver (below right, *Castoroides ohioensis* from Clyde, New York, USA) and probiscideans (bottom, *Gomphotherium* sp.).

percentages. Many more members of the megafauna, meiofauna and microfauna shifted their geographic ranges at this time. Extinctions of marine invertebrates also occurred in the Pleistocene, most notably in the western Atlantic and Caribbean (see p.166). Prominent victims of the Pleistocene extinction are shown below.

Although these extinction rates look impressive, Phillip Gingerich has shown that they are not atypical of values for other Caenozoic series (Palaeocene, Eocene, Oligocene, Miocene, Pliocene). Moreover, Anthony Barnosky and colleagues argued convincingly that even the megafaunal focus of the Pleistocene extinctions is not surprising insofar as species characterized by large body sizes also tend to have small population sizes, small taxonomic richnesses, are ecological specialists, require large geographic ranges to sustain themselves, and have relatively long generation times during which they produce few offspring. These are all characteristics that increase overall extinction susceptibility.

BELOW Prominent victims of the Neogene extinctions: giant ground sloth (below left, *Nothrotheriops*); hominid primate (below right, oblique view of an adult female *Neanderthal* cranium discovered at Forbes Quarry, Gibraltar) and giraffid (bottom, *Sivatherium giganteum* from Pinjor, India).

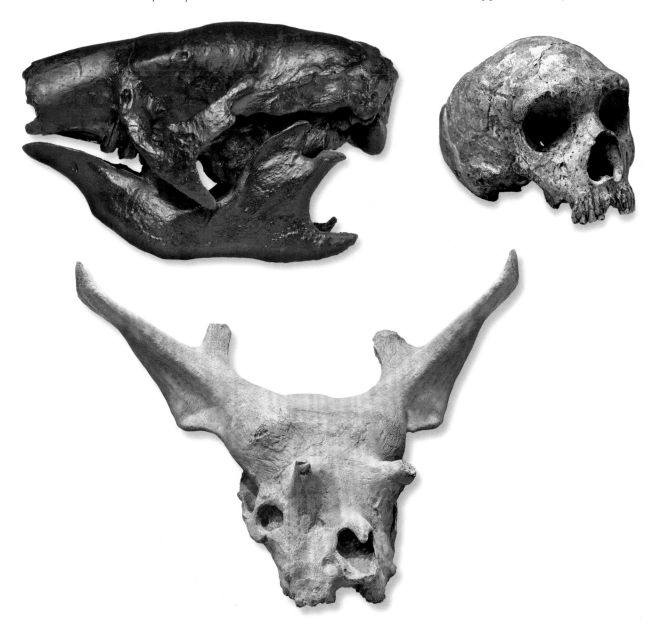

Prominent victims of the Neogene extinctions.

ABOVE giant goose, *Dromornis stirtoni*, with chicks, Australia.

ABOVE RIGHT Pliocene terrestrial environment, South Africa. Foreground: black-backed jackal (left, *Canis mesomelas*), crowned eagle (centre, *Stephanoeatus* sp.), sabre-toothed cat, *Homoterium* sp.

MIDDLE a group of giant wombat, *Diprotodon*. The hippopatamus-sized marsupials were one of the largest marsupials to have lived.

RIGHT A northern Siberian Late Pleistocene terrestrial environment 1.8 million years to 11,000 years ago, with wolverine, *Gulo gulo*, woolly mammoth, *Mammuthus primigenius*, and horses, *Equus* sp.

BELOW Two giant kangaroos, *Procoptodon*, which were restricted to the Pleistocene of Australia.

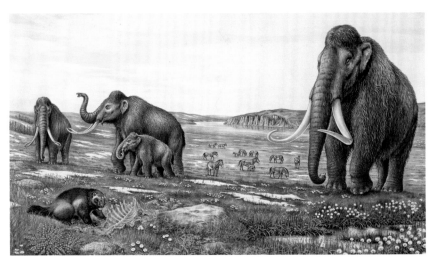

LEFT A giant ground sloth, *Megatherium americanum*, from South America.

BELOW An artist's reconstruction of a chalicothere, *Chalicotherium grande*, a mammal with very short hind legs that walked on the knuckles of its forefeet.

BOTTOM Skeleton of an extinct sabre-toothed cat, *Smilodon* sp., which lived about 15,000 years ago in North America. It was about the size of a present day lion.

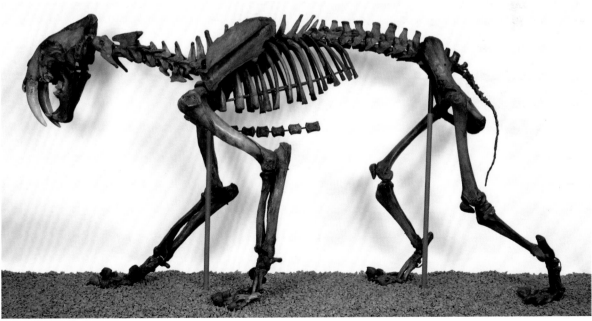

TIMING

In North America, South America and Europe the megafaunal extinctions are clustered at *c.* 11,000 years and took about 10,000 years, though in all three cases regional extinctions and changes in geographic range occur much earlier. In Australia the megafaunal extinctions began much earlier, perhaps as long as 30,000 years ago. The chronology of Pleistocene extinctions in Asia and Africa has yet to be established.

CAUSE(S)

Two primary causes have been advanced to explain the megafaunal extinctions: climate change and human hunting. The climate change hypothesis points to palaeobotanical and palaeotemperature analysis that demonstrate a correlation between physical and biotic changes in local environments and the disappearance of large herbivores and carnivores. It also points to the megafaunal survivors as being an important source of data.

Several authors have noted the mammals that continued into the subsequent Holocene are characterized by feeding strategies that allow them to survive in landscapes characterized by low diversities of plant material including plants that are toxic to other mammals. Ruminants such as bison, deer, moose and sheep, for example, accomplish this through use of multiple stomachs and complex food-processing procedures. Large herbivores that lack such adaptations to poor forage (e.g. proboscideans, horses, camels, sloths, peccaries) were much more likely to become extinct, either regionally or globally, depending on the geographic distribution of the species. Of course, any carnivores that specialized in hunting particular types of herbivores – which is to say most large carnivores – would have the same extinction susceptibilities similar to their prey. Proponents of the climate change scenario also point out that herbivores are linked to the composition of plant communities through co-adaptation via patterns of botanical succession. Once herbivorous species that play critical roles in maintaining some part of botanical diversity within a region are gone that diversity declines, setting up a negative ecological feedback loop. Finally, the climate change hypothesis is the only causal explanation that has been credibly advanced to account for patterns of extinction among smaller terrestrial vertebrates and marine invertebrates.

Critics of the climate change hypothesis argue that ecological analogues, and in some case the same species, had survived similar episodes of climate change in the Late Pliocene and Early Pleistocene. Additionally, they note that in some cases (e.g. horses) species have been reintroduced into areas from which they were eliminated during the Pleistocene and have thrived, though it should be noted that the reintroduction of a single species is not the same as re-establishment of an entire ecology that, in its original state, included competitors that are extinct (e.g. ground sloths, mammoths, mastodons, woolly rhinoceros).

The competing causal scenario pertains only to the megafaunal component of the Pleistocene extinctions and argues that the introduction of human populations to areas from which humans had been excluded (e.g. North America, South America) resulted in the elimination of large mammalian herbivores as a result of habitat destruction and hunting (Haynes 2003). This scenario is sometimes referred to as 'Pleistocene overkill', or the 'overkill hypothesis'. Proponents point to the close correspondence between the date of the megafaunal extinctions in North America and the dates obtained from the oldest human campsites and artefacts discovered on that continent. They also note the fact that human artefacts have been found associated with the remains of five extinct megafauna species in the North American Pleistocene (mastodon, mammoth, camels, horses, giant tortoise) and argue that large, slow-moving herbivores are precisely the animals ancient humans would have chosen to hunt. Critics of this scenario point out that, while the coincidence between human migration and megafauna extinction dates are indeed convincing in the case of North America, the evidence for a link between human migration and extinctions in other parts of the world is either absent or far less compelling. These critics also note that human hunting cannot be accepted as a credible explanation for all the Pleistocene extinctions (e.g. non-megafaunal extinctions), and that, since the ability of humans to migrate into North America was made possible through climate change (e.g. exposure of the Bering land bridge from beneath glacial cover), human hunting, at best, represents an interesting proximal - but not the ultimate - cause for a limited number of large herbivore extinctions in a single region of the globe.

Despite extreme statements on both sides of the Pleistocene extinction debate it seems clear that both mechanisms were operating in this interval. Owing to the inherently limited nature of the human hunting scenario, it also seems clear that climate change is the dominant driver for most of the biotic phenomena – global extinctions, regional extinctions, changes in biogeography in both terrestrial and marine habitats – seen in the Pleistocene fossil record, including the migration of human populations into North America. Although human hunting contributed to this extinction event, to me there seems little reason to believe human hunting alone was responsible for the Pleistocene megafaunal extinction worldwide and at all taxonomic levels. There is also little reason to believe that extinctions which clearly could not have been caused by humans are in any way less important to our understanding of this interval in Earth history than those that were.

&. Given   by G. EDWARDS

14 The modern and future extinctions

CCORDING TO FORMER US VICE PRESIDENT Al Gore's influential book, the world loses approximately 40,000 species a year to extinction. The popular media abounds with stories that half the world's species will become extinct in the next 100 years. As we have seen, the idea that 50% of extant species could go extinct, either as a result of a single cause or a coincidental concatenation of causes, is not only credible but would only qualify as one of the lesser extinctions known to have taken place over the course of Earth history. Extinction events of this magnitude are tolerably common in the fossil record and there is no good reason to suspect they cannot – or will not – happen again. The century-long timeframe of these estimates is, admittedly, on the short side geologically speaking. But some students of the fossil record have suggested much larger ancient extinction events have occurred over even shorter timeframes. How can we estimate future extinction rates and how reliable are the estimates that we see cited in the newspapers and magazines and hear discussed on the radio and television? Most importantly, is there anything we can, or should, do to prevent an extinction caused by the usurpation of the planet by a single species – *Homo sapiens*.

MODERN EXTINCTION RATES

A number of compilations of species loss since the end of the Pleistocene (11,700 years ago) exist. These usually contain less than a thousand species, some much less. Modern extinction lists inevitably include species that are familiar icons of extinction, such as the passenger pigeon (*Ecopistes migratorius*) and the dodo (*Rhaphus cucullatus*), that have occurred during the course of human history. But most of the species included in these lists are quite obscure, despite their wonderfully evocative names (e.g. the guayacon ojiazul, the New Caledonia lorikeet, the toolache wallaby). While these lists surely underestimate the true magnitude of Holocene extinctions and can only claim any accuracy in the interval since systematic written tabulation of extinct species began in the seventeenth century, this estimate of *c.* 1,000 confirmed extinctions or known species represents a percentage extinction intensity of *c.* 0.05. In addition, it should be appreciated that the taxa comprising these lists are dominantly island species (*c.* 75%) of terrestrial gastropods, insects and birds, none of which has a particularly rich fossil record.

Looking at these data in another way, Table 1 listed average longevities for different taxonomic groups as estimated from the fossil record. Based on these data we can estimate an average longevity over all species with a medium to high fossilization potential. Depending on the specifics of the estimation method and composition of the sample this figure ranges from 5 to 10 million years, which translates roughly into one species per year. Drawing on this result, since 1600 we should have witnessed the extinction of *c.* 400 species as a consequence of 'natural processes'. This means the historical extinction rate, which significantly underestimates the true extinction rate, exceeds the palaeontologically estimated background extinction rate by at least a factor of 2. Adjusting the total number of extinctions for fossilization potential diminishes the difference between observed and expected extinction numbers, but does not account for the whole of this discrepancy. There are three conclusions we can draw from this simple exercise: (1) the observed extinction rate is well above what would be considered 'natural' based on the data of the fossil record, (2) the average historical rate has exceeded the natural background rate for the whole of the last 400 years and probably well beyond, and (3) the average historical extinction rate is nowhere near as large as would be expected over the course of one of the 'Big Five' mass-extinction events. Indeed, based on the elimination of higher taxonomic groups, it is less than the rates observed during various minor extinction events such as those of the Miocene, Pliocene and Pleistocene.

So, where do the claims that the Earth is currently in the midst of a sixth mass extinction (see Leakey 1996) come from? Without exception such assertions come from estimates based on extrapolations or scalings from small, localized datasets collected over small time intervals, usually from regions and involving taxa not well represented in the fossil record.

Some of the more outlandish claims for the modern extinction record can be dimissed easily. Take Gore's claim that 40,000 species are going extinct each year, for instance. This figure comes from a 1979 book, *The Sinking Ark: A New Look at the Problem of Disappearing Species,* in which its author, Norman Myers, asks the reader to simply accept that 1 million species have become extinct from 1975 to 2000. This assumption implies an average extinction rate of 40,000 species per year. But Myers cites no data in support of his claim, simply stating it is a 'reasonable' estimate. Since only around 1.7 million living species have been described thus far, Myer's prediction is either a very poor estimate of the actual modern extinction rate or the total number of species present on Earth is vastly greater than any biodiversity researcher has offered to date.

Other claims are based on data, many compellingly so. For example, in 1995 David Steadman published results of a study of Holocene bird extinctions on Pacific islands, contrasting extinction rates in the interval prior to and after colonization by Polynesian peoples. In three broad regions (remote outpost islands, the Polynesian heartland, Micronesia–Melanesia), bird faunas recovered from the zooarchaeological records were compared. In each region the number of species disappearing from the

Birds B.M.Vol.XVIII.   Pl.

Cabalus dieffenbachii .Juv.

LEFT Hawkin's Rail, a recently extinct flightless bird indigenous to New Zealand.

local fossil records of the islands studied increased for post-colonization samples, in most cases dramatically so. Largely, the victims of this extinction were flightless birds – especially rails (see above). Owing to their fully terrestrial nature, rails had evolved into separate species on (virtually) each island. Moreover, being small and flightless, they would have been relatively easy for humans and other introduced predators to catch. In Steadman's study the lowest extinction rates were recorded from seabirds that fly from island to island and hunt in the open ocean waters.

Based on these local records from a sample of islands in each region Steadman estimated an extinction/extirpation rate of 10 species or local populations on each of 800 major islands though only anecdotal statistics were provided to support this estimate. Taken at face value these data suggest that as many as 8,000 local populations may have been lost as a result of human colonization, through a combination of direct hunting, habitat loss as a result of human occupation, the introduction of new predators or competitors that came to each island as a result of human discovery, and/or the inadvertent introduction of pathogens. Steadman then observed that, since by his estimate 25% of the avifauna on each island constituted flightless rails, the estimated loss of 8,000 populations suggests the extinction of as many as 2,000 rail species alone.

Obviously Steadman's results indicate a much higher Holocene extinction rate among island birds than is included in the tables of species known to have become extinct over the past 400 years, but also a much lower rate than asserted by Myers and accepted by Gore. Steadman's rate is also much higher than the number of all species extinctions known from the Holocene records of the major continents. His study makes a convincing case for the association between human colonization of the islands and species extinctions. However, many researchers believe it is misleading to compare these types of data to those derived from studies of the fossil record directly. In a sense such studies demonstrate only that the level of detail of data collected from most modern species lies well above the resolution afforded by the fossil record. Only four Pleistocene bird genera are present in the Sepkoski genus dataset and each is still extant. Islands were no doubt present at all times in the ancient past. But like lakes and rivers, islands are short-lived structures.

The oldest modern island is Mangaia, one of the Cook Islands, which is dated at just over 18 million years. The oldest of the Hawaiian islands – arguably the island chain whose fauna and flora have been studied most intensively – Kauai, has existed for 5.0 million years. The big island of Hawaii is a mere one million years old.

There is no doubt islands occurred in ocean basins in the geological past and were populated by animal and plant species, many of which would have evolved over time into indigenous species. All the species that existed on all ancient islands are part of the potential extinction record. Because of their small size and short duration as identifiable features in the stratigraphic record though, island biotas make little contribution to the palaeontological datasets used to study ancient extinctions. As a result, it is very difficult to compare data derived from sources and extrapolations of the sort used by Steadman with the data of the ancient fossil record and to use results of such comparisons to argue for a similarity in extinction magnitude. One (statistical) way out of this dilemma is to introduce taxonomic scaling into the calculations and ask not how many species were eliminated on islands during the Holocene, but how many genera or families were eliminated globally as a result of island extinctions. The data provided in Steadman's article suggests that if taxonomic scaling is included as a factor to render the data comparable to compilations derived from the fossil record the ancient and modern extinction records come into much closer agreement.

COMPARING THE PRESENT WITH THE PAST

As the debates about modern extinction rates mature it is becoming increasingly evident that, for a number of reasons, it is inappropriate to simply scale up extinction data derived from limited studies of modern species and compare the results to the great extinctions of the geological past. Attempts to do so miss the point of extinction studies in both the modern and ancient realms. As noted by Geerat Vermeij (2004) the extinctions we see in the modern world are essentially local in character, driven by factors such as habitat fragmentation, the introduction of predators or competitors, and the introduction of pathogens. These factors can be serious, but they fall far short of causing the elimination of substantial numbers of species over entire regions or precipitating extinction cascades. No doubt such factors played a role in ancient environments as well. But these factors are not among the causes of the great palaeontological extinction events. Rather, these are caused by the elimination of entire categories of habitat (e.g. complete disappearance of the shallow marine floor of a large epicontinental sea as a result of a 100 m or 330 ft drop in sea level), global refrigeration of the planet and/or the sudden disruption

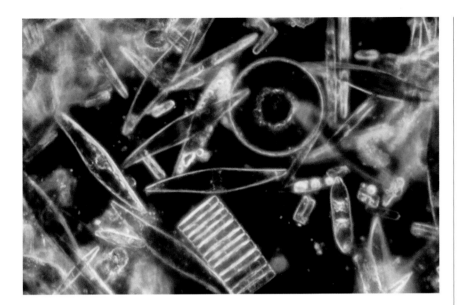

LEFT Diatoms are one of the most common types of phytoplankton and, as such form the basis for many marine and freshwater food webs. The great extinctions of the geological past, especially the largest or mass extinctions, are often associated with evidence of a decrease in the diversity of phytoplankton.

in the primary productivity of the oceans. These far more powerful and intense processes operate on regional and global scales, cause the ecological collapse of major biomes (e.g. reefs) with ensuing extinction cascades, causing whole categories of biodiversity to be eliminated.

In a sense the difference between the present and the past in terms of extinction is the difference between consumers and producers. So far, modern extinctions have primarily affected ecological consumers whose loss, regrettable as it is, rarely perturbs other species living in the same habitat. Ancient extinctions – especially the great extinctions – affected the habitats of primary producers on whose existence all species depend to a greater or lesser extent. If these features of the biosphere go, even for a short time, they will not be able to be overlooked or relegated to a few column inches in the local newspaper. The current challenge is to determine whether, and when, levels of modern species extinction are likely to become dangerous and balancing decisions that must be taken to prioritize economic, research and educational resources on the most important problems facing human populations at any given time. To do this scientists, regulators and politicians will require the best available estimates of predicted rates, consequences and strategies to avoid species loss.

ESTIMATING FUTURE EXTINCTION RATES

At present three methods are used to estimate future extinction rates: species-area relations, extrapolations from the so-called 'red lists' of threatened and endangered species published by the International Union for Conservation of Nature (IUCN), and the probability based estimation procedure, which also uses the IUCN red lists as a source of data.

The species-area effect was first described by Robert MacArthur and Edward O. Wilson and states that a regular mathematical relation exists between the size of islands and the number of species they contain. Mathematically, this relation is usually expressed in the following manner:

$$S = cA^z$$

where: S = number of species

A = habitat area

c = a constant that expresses the value of the species richness in the smallest sampling area (= y-intercept)

z = the rate of change in species number as a function of an increase in area (typically in the range of 0.2–0.3 for islands).

May (1995) points out that, in terms of extinction estimates, if we assume the change in habitat area is large ($\triangle A<<1.0$) we can rewrite this equation to express proportional changes as follows.

$$S=z \quad A$$

where: S = proportion loss/gain in species numbers,

A = proportion gain/loss in habitat area.

With this relation in hand we are in a position to estimate the magnitude of species loss we can expect under various scenarios of environmental change. For example, in both Europe and North America the area of primary forest has diminished by 98–99% from its state at the beginning of the nineteenth century. If we assume a typical island value of 0.25 for z, our estimate of the proportion of species loss would be c. 25% of extant species. As there are about 250 forest bird species in North America this would suggest that we should have seen the extinction of some 60 species over this 200-year period. Unfortunately, this estimate does not conform to the actual number of recorded extinctions within this interval, which is three species. Jared Diamond and Robert May point out that the original values of z were inferred using data from physical or ecological islands and have suggested that, in order to accurately reflect the character of extinction data on continents, z must be re-estimated using continental data. However, even their suggestion of dropping the z-value to the range of 0.1–0.2 does not accurately predict the anomalously low extinction levels seen among North American or European forest birds.

In order to address the problems associated with the raw estimation of species reductions from habitat loss, and to acknowledge the uncertain time span over which species extinctions actually take place, many conservationists now advocate interpreting species-area results as indicating the proportion of species 'committed to extinction' over some (unspecified) timeframe in the sense that wild populations are regarded as being no longer viable unless conservation actions are taken. In 1992 V. H. Heywood and S. N. Stuart used this version of the species-area relation to argue that 450 bird species would be committed to extinction by 2015. While this figure seems quite large it still represents less than 5% of extant bird species. It is 2012 now and this book will appear sometime in 2013. I very much doubt that a few years hence researchers will be able to present evidence that 450 bird species became extinct in the interval between 1992 and 2015.

Recently Fangliang He and Stephen P. Hubbell have challenged the 'committed to extinction' interpretation of species-area data, arguing that 'the area required to remove the last individual of a species (extinction) is larger, almost always much larger, than the sample area needed to encounter the first individual of a species, irrespective of species distribution and spatial scale' (He and Hubbell 2011, p.68,). This research suggests that the problems of estimating the correct number of species present in an area and estimating the number of species extinct are different and require different mathematical models to resolve correctly. In particular, they argue that even if all other factors are held constant it requires a much greater loss of habitable area to drive a species extinct than it takes an increase in habitable area to support a new species. The implications of their finding is that most, if not all, previous estimates of species loss due to a reduction in habitat area have grossly overestimated the amount of species loss that will take place per unit habitat area lost, a finding that accords with the empirical data on actual species losses. This research will take time to confirm, but it and other research studies like it should improve the species extinction estimates that can be made using species-area relation models. The second most common way to estimate future extinction rates has been to examine the histories of threatened species designation on the IUCN red lists, which uses seven categories to assess the state of a species in relation to extinction (see Table 4).

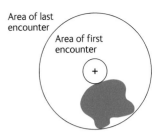

ABOVE He and Hubbell's 2011 model of the species-area effect showing that for any arbitrary starting point (+) or sampling frame (here circular) the area that must be searched to identify the presence of a species is always much smaller than the search area required to confirm its absence.

Table 4. Species threat categories used by the IUCN to assess extinction risk.

Category		Definition
Least concern		An extant taxon which has been evaluated but does not qualify for any other IUCN category
Near threatened		A taxon that is considered threatened with extinction in the near future, although it does not currently qualify for any of the Threatened categories; a taxon whose extinction status should be re-evaluated often or at an appropriate interval
Threatened categories	Vulnerable	A taxon likely to become Endangered unless the circumstances threatening its survival and reproduction improve; vulnerable species are monitored
	Endangered	A taxon which is at risk of becoming extinct because it is either few in numbers, or threatened by changing environmental or predation parameters
	Critically endangered	A taxon in which numbers have decreased, or will decrease, by 80% within three generations
Extinct in the wild		A taxon whose only known living members are being kept in captivity or as a naturalized population outside its historic range
Extinct		A taxon in which no living members are known

These categories have changed over time as research improves and more information becomes available for individual species.

The Red List Method assigns a number to each category, for example 0 for Least Concern or Near Threatened, 1 for Vulnerable, 2 for Endangered, and tabulates the direction and amount of movement within the categories for each species. From this a median change in status for any taxon can be computed, a number that, in all cases, is quite small for almost all major organismal groups because most species are not included on the IUCN list. Given these data, the time to extinction for any group can be estimated by calculating the ratio of the difference between its current position and the median change in its status. This approach to estimating extinction risk assumes that the data collected for each species are accurate, that the species has been sampled over a sufficiently long time to even out short-term spikes in the status change estimate, and that any conservation efforts that might be applied to mitigate population loss will be ineffective.

The Red List Index method cannot be used for groups most of whose members have not been assessed by the IUCN (and so include representatives of a large number of '0' taxa by definition rather than as a result of assessment, e.g. insects). But for well-studied groups (mammals, birds, amphibians) in which most species have been allocated a position in the classification as a result of active assessment the index provides a crude estimate of extinction risk. This method is far from perfect, but it is better than nothing. Using this method it has been estimated that half of extant birds and mammals will likely become extinct within 200–300 years. Reliable figures cannot be obtained for any other group using this method at present owing to lack of sufficient IUCN resources.

A third method is even more restricting in terms of the number of taxonomic groups that can be considered because it requires the development of probability distributions for extinction as a function of time for each species. This is a more accurate approach than the use of the raw IUCN data, but requires an even larger investment of time and energy by investigators. When these data are used for orders and families of birds, reptiles and mammals the estimates of the time required to eliminate half of extant high taxa range from 100 to 1,000 years with a frequency peak of estimates for mammals and birds between 300 and 400 years.

For example, the plants and animals of Great Britain are unquestionably the best studied in the world. Using data from the British Red Books, entomologists N.A. Mawdsley and N.E. Stork were able to show that, since 1900, less than 1% of all British insects suffered local extinction and none became extinct globally. Currently, about 6% of British insects are considered threatened. Mawdsley and Stork report similar results for British marine invertebrates, data that, if anything are more comparable to those of the fossil record than any others discussed in this chapter.

RESPONSIBILITIES

We can draw several general conclusions from this brief review of the state of future extinction estimates. The first, and arguably the most important, is that our understanding of how to estimate future extinction rates and probabilities is at a very rudimentary state of sophistication. Coupled with this, while the data we have to work with are good in parts for some charismatic groups (e.g. mammals, birds), they are woefully inadequate to non-existent for most of the species — even many of the charismatic species — we are interested in protecting. Much more research and much more thought will be needed for science to deliver on its promise to supply the commercial sector, government, and the general public with accurate and objective data on extinction risk uncoloured by unduly alarmist, and ultimately unsupportable claims.

But rather than closing this discussion with a blanket plea for increased resesarch funding, allow me to acknowledge frankly the obvious fact that resources are limited and, in a very real sense, they will never be sufficient to the task of acquiring the information needed to make the decisions that must be made within the required timeframes. In this sense science must get much better at playing its role in determining the most efficient use of the resources that can be made available. It is up to research communities and major scientific research institutions to set informed priorities for themselves with the goal of providing the data, analyses, and predictions that society needs address its top concerns irrespective of whether those are the top personal concerns of individual scientists.

15 Summary and conclusions

EXTINCTION STUDIES HAVE ALWAYS BEEN, and probably always will be, seen by some as the shadowy negative partner of evolution studies. Whereas evolution is regarded in a positive light as being about discovery, origins, potential, novelty and life, extinction is typically regarded as being about limitations, endings, constraint, exhaustion and death. But just as death is an inextricable part of life, so extinction is a fundamental aspect of evolution. Without extinction evolution would quickly grind to a halt.

OPPOSITE One of the over 140 frog species known from Sri Lanka. Many amphibian populations are in decline worldwide as a result of disease, habitat destruction, habitat modification, exploitation (hunting), pollution, pesticide use, introduced species, and increased ultraviolet-B radiation (UV-B).

RECAPITULATION

Science has been rather slow to recognize extinction's significance. Speculations about organic evolution can be traced to ancient Greek, Roman and Chinese philosophy. However, extinction was not discussed seriously in the biological literature until the 1750s and was not acknowledged as a real phenomenon until Cuvier's palaeontological investigations at the turn of the eighteenth century. Cuvier regarded extinction as being brought about by a series of natural revolutions; environmental events whose intensity was unprecedented in recorded human history. But his 'catastrophist' view of Earth history was opposed by Charles Lyell, Charles Darwin and other 'uniformitarians' who preferred to see both evolution and extinction as part of the normal operation of the world. Specifically, the uniformitarians (or actualists) regarded extinction as attributable to processes that go on continuously in nature and that can be studied by anyone. Uniformitarian researchers explained the odd juxtapositions of extinct species that supplied the empirical evidence for the catastrophist view as misleading imperfections in the fossil record.

Generations of geologists were taught the uniformitarian view of Earth history between 1900 and the mid-1960s. Ironically though, over this same interval the intellectual descendants of Lyell and Darwin were gathering empirical evidence from the fossil record that would eventually to overturn the simplistic uniformitarian model; evidence that several major disruptions in the continuity of evolution had indeed occurred in deep time.

CATASTROPHISM REVISITED

By 1960s prominent palaeontologists such as Otto Schindewolf in Europe and Norman D. Newell in the US were advocating the recognition of six 'mass-extinction' events each of which occurred at the ends of major geological periods: end-Cambrian, end-Ordovician, end-Devonian, end-Permian, end-Triassic and end-Cretaceous. It was thought these events were driven primarily by sea-level change and took millions of years to complete. In this way palaeontologists sought to reconcile the facts of the fossil record with the gradualist/actualistic philosophy on which Darwinian evolution was grounded.

Between the mid-1960s and 1980 palaeontologists became interested in looking at major patterns in the stratigraphic distribution of fossil species that previous generations of their colleagues had spent their professional lives amassing for the purpose of dating sedimentary rocks. In 1982 the University of Chicago palaeontologists David Raup and Jack Sepkoski used a dataset that included the stage/series-level stratigraphic ranges of over 2,800 families of fossils to refine Newell's analysis. Raup and Sepkoski recognized five of Newell's six events as unusually large extinction events since their family-level data did not show the end-Cambrian extinction to have been particularly large. Moreover, they argued that recognition of a distinction between 'mass extinctions' and 'background extinctions' is fundamental to the development of a complete understanding of evolutionary processes. Under this model extinctions occurring during background extinction intervals are generated by normal Darwinian selection pressures, whereas mass extinctions represented time intervals during which a set of processes operated that were outside the domain of normal natural selection. Owing to the magnitude of, and time intervals between, mass extinction events survival was not due to superior adaptations that had been gained by species through exposure to background extinction events, but was more akin to a random winnowing. As a result, mass extinctions were held to have the ability to 'reset' the evolutionary clock by extinguishing seemingly well-adapted lineages and allowing others to take their ecological places. Raup and Sepkoski followed their original analysis of family-level data with an analysis of genus-level data which revealed largely the same patterns.

Also during the late 1970s and continuing throughout the 1980s an interdisciplinary team of physical scientists led by Luis and Walter Alvarez recovered evidence that the Earth had been struck by a large extraterrestrial bolide at a time coincident with the boundary between the Cretaceous and Palaeogene periods, precisely the point where John Phillips, Norman Newell, Dave Raup, Jack Sepkoski and others had located their end-Cretaceous mass extinction. This result was seen by many as legitimizing Cuvier's catastrophism in the sense that (1) no bolide impact of similar magnitude had occurred in recorded human history and (2) extinction patterns across the Cretaceous–Palaeogene boundary seemed to have occurred suddenly and to conform to a random survivorship model, at least for some organismal groups (e.g. bivalve and gastropod molluscs). Based on this juxtaposition between the physical evidence for bolide impact and the biotic evidence for mass extinction some enthusiasts speculated that bolide impact might provide general explanation for the phenomenon of mass extinction — the Single Cause (SC scenario) of mass extinction. However, research undertaken since the mid-1980s has not borne out many of these new ideas regarding the causes of large extinction events or their role in mediating evolutionary processes.

THE POST-MODERN SYNTHESIS

Analysis of the magnitudes of stage/series-level extinction events through the Phanerozoic indicates that the set of extinction intensities forms a single, continuous distribution. There is no objective basis on which to recognize 'mass extinctions' as being caused by a different class of processes in the sense that any class of processes can claim to be associated uniquely with large losses of species over stage/series-level timescales. Indeed, when compared with one another using the same metric the 'Big Five' mass extinctions can neither be separated from background extinction events by noticeable discontinuity in extinction intensities nor constitute the five largest extinction events. A more reasonable interpretation of these data is that these 'Big Five' extinction events are history-driven (rather than process-driven) anomalies and represent time intervals during which unusually intense manifestations of the same processes that operate during normal background extinction intervals occur together coincidentally — the Multiple Interacting Causes (MIC scenario) of mass extinction. While amplification of the effects of normal selection processes through historical coincidence may have the same macroevolutionary effect as a single cause mass-extinction processes (e.g. when taken as a whole patterns of survivorship across different organismal groups through such an interval may be more a by-product of random chance than prior, selection-mediated adaptation), the difference between these causal modes matters greatly in terms of our understanding of evolution for one (SC) is fundamentally ahistoric in character whereas the other (MIC) is generated by the very details of environmental change history.

A survey of the major extinction events throughout the Phanerozoic (Table 5, p.190) shows that each took place over an extended time interval relative to the duration of ecological and evolutionary processes and each took place over a time interval during which multiple factors, each of which represent normal players on the Earth's environmental stage, were interacting coincidentally to bring about unusually large, complexly structured, and long-lasting intervals of environmental instability. To be sure, unique ahistorical processes such as bolide impact overlay this pattern of coincidental alignment of Earth-generated processes of environmental change. No doubt these ahistorical processes are responsible for some proportion of the disappearances observed during so-called 'mass extinction' intervals. But there is exceedingly little evidence that any single cause was responsible for the great extinctions of Earth's ancient past.

Table 5. Proximate and inferred ultimate causes of major Phanerozoic extinction events.

Extinction event		Causes
Late Cambrian	Proximate	Sea-level fall, sea-level rise, marine anoxia
	Ultimate	Tectonic (spreading rate changes associated with opening of Iapetus Ocean); Environmental state (some glaciation (?), stagnation of marine circulation patterns caused by development of a strong thermocline via climatic induction)
End-Ordovician	Proximate	Phase 1: global cooling, sea-level fall, increased marine circulation, ventilation of marine deep waters, marine nutrient sequestration. Phase 2: global warming, sea-level rise, decreased marine circulation, changes in marine nutrient flux, stagnation of deep oceans, spread of dysoxic and anoxic waters over continental shelf habitats.
	Ultimate	Tectonic (movement of Gondwana over the southern pole initiating a deep glacial episode and sudden termination of that episode after a geologically short, but ecologically long, interval [c. 1 million years]; Environmental (stagnation of deep ocean waters)
Late Devonian	Proximate	Phase 1 (Frasnian) sea-level rise, marine anoxia, global warming followed by global cooling induced by removal of atmospheric CO_2. Phase 2 (Famennian) glaciation, sea-level fall, global cooling
	Ultimate	Late Frasnian-Tectonic (spreading rate induction of sea-level rise; LIP volcanism [Vilnuy Volcanic Province eruptions], Late Famennian-Tectonic (global cooling caused by continental glaciation in response to Gondwana moving over the South Pole)
Late Permian	Proximate	Middle Permian – sea-level fall, global cooling. Late Permian – sea-level rise, extreme continentality, global warming, marine anoxia, marine nutrient sequestration, collapse of primary productivity.
	Ultimate	Tectonic (southern Pangea (= ancient Gondwana) moving over South Pole, extreme continentality caused by continental drift; LIP volcanism; Environmental (oxidation of Gondwana coal deposits)
End-Triassic	Proximate	Sea-level fall, sea-level rise, marine anoxia, global cooling
	Ultimate	Tectonic (inferred change in mid-ocean ridge heat flow), LIP volcanism
End-Cretaceous	Proximate	Sea-level fall, habitat fragmentation, species-area effect, reduction in primary productivity.
	Ultimate	Tectonic (inferred change in mid-ocean ridge heat flow), LIP volcanism (Deccan Trap eruptions), bolide impact (Chicxulub).

This view of major extinction events in Earth history unites Newell's original suspicion that habitat destruction brought about as a result of a combination of sea level and climate change were the mechanisms that induce most major extinction events, with more recent research demonstrating the variety of mechanisms can produce these changes. When these mechanisms operate singly and/or at low intensities over the course of a stratigraphic stage/series, minor or background extinction rates result. However, when two or more of these ultimate environmental change mechanisms come into coincidental alignment and/or operate at high intensities, major disruptions to patterns of evolutionary diversification can occur, disruptions that can have far-reaching consequences.

In final support of this MIC interpretation of the great palaeontological extinction events it is noteworthy that, over the last 250 million years – the time interval for which geologists have the best data – there have been three major (end-Permian, end-Triassic, end-Cretaceous) and seven minor (Pliensbachian, Callovian, Tithonian, Aptian, Cenomanian, Late Eocene, Middle Miocene) peaks in extinction intensity (see below). The impact crater record suggests that medium

BELOW Comparison of the stage/series-level stratigraphical distribution of short-term sea-level falls, continental flood-basalt volcanic events, and bolide impact events with Phanerozoic extinction intensity. The small number alongside the bolide-impact icons represents multiple impacts of this size in the associated stage/series. Bolide icon sizes represent crater diameters: small icons = 1–10 km; intermediate icons = 10–100 km; large icon = >100 km.

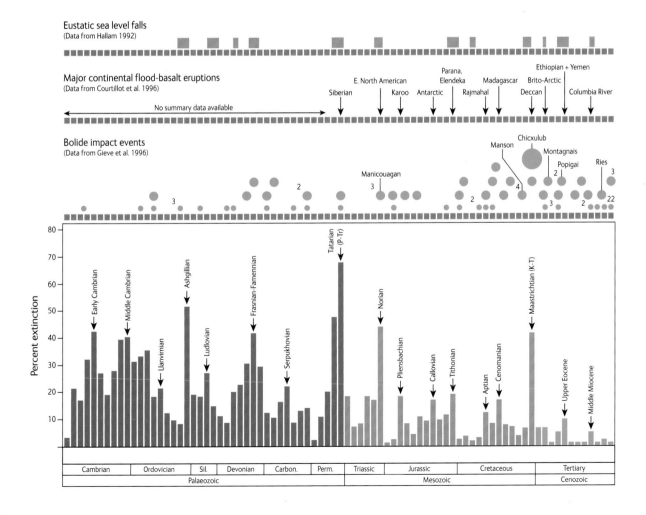

to large-sized impacts occur in 23 of the 42 stages. Of these 23 stages whose environments were perturbed by sizeable bolide impacts, only five coincide with extinction-intensity peaks and only one with a 'Big Five' mass extinction. Thirteen of these 42 stages contain evidence of substantial sea-level regressions. But of these only six coincide with a major or minor extinction-intensity peak. However, 11 non-oceanic LIP volcanic eruptions have occurred in this same interval. Of these, fully 10 coincide with an extinction-intensity peak when the data are parsed at the stage level including all three of the "Big Five" mass extinction events. Perhaps most importantly, all three of the largest extinctions to have occurred during this 250 million year interval coincide with an LIP volcanic event a sea level regression while only one – the end-Cretaceous event – adds a large bolide impact to the panoply of putative causal mechanisms. The statistical association between the SC and MIC scenarios has been investigated (MacLeod 2005) via Monte Carlo simulation. Among the single mechanisms only LIP volcanism exhibits a statistically significant association with extinction peaks. The highest statistical significance, however, was obtained for the MIC scenario.

LESSONS LEARNED

Given the fact that extinctions both large and small have occurred throughout the Earth's geological past, it is only natural to ask what lessons can be drawn from this past that might inform our future management of the planet, particularly as the spectre of widespread species loss as a direct result of human activities looms before us all. The historical record of documented species losses has not yet reached the point at which the modern extinction event can be credibly compared to even middle-sized background events from deep time. Nevertheless, this discrepancy should not be a cause for comfort. The documented historical record surely underestimates the true scale of species extinctions over the last 400 years by several orders of magnitude. In addition, various techniques for estimating future extinction numbers from current extirpation trends suggest that biodiversity loss among currently extant species could reach geological proportions in around 100–500 years, assuming present trends continue. And therein lies the real problem.

The data from ancient and modern extinctions are clear. Major extinctions happen when a set of causal factors that might not be of serious consequence by themselves become aligned in time. Not only are we living in such a time (e.g. sea level currently stands as low as it has at virtually any other time in Earth history and sea-level lowstands are known to amplify the extinction-related effects of other mechanisms), humans are now, collectively, a factor of environmental change fully as significant as a large LIP eruption and/or bolide impact; perhaps larger. In the geological past the application of human intelligence, technology and social cooperation probably contributed to aspects of the Pleistocene megafauna extinctions. Humans have unquestionably been responsible for the decimation of bird species on oceanic islands over the last 400 years. Since the Industrial Revolution, and certainly over the last 50–75 years, our technological skill and sheer numbers have reached the point where our species is posing a credible threat to the biota of entire continents and to the seas themselves. However, our best evidence suggests that the resilience of life is such that species losses will continue to be the exception rather than the rule for the immediate future. Over time these losses will mount up, possibly at a much higher rate than they did during the great extinctions of the past. Nevertheless, on a day-to-day basis the Earth's environments will likely appear stable to the casual observer, with changes only becoming apparent to those who can think back over decades and remember what the environment – what the world – used to be like. Those who lack this personal historical perspective may fail to perceive species extinction as a serious threat and by the time they do it may be too late to save a sufficient number species (especially primary producers) to avoid extinction cascades. Educational efforts can be mounted to correct this misperception. But there are many problems that beset human societies and the ability of societies to cope with them all is, and will continue to be, limited. In this way, problems that play out over long time spans will always be at a disadvantage relative to problems that strike close to home and so demand more immediate human attention.

Still, there remains cause for hope. For virtually the whole of the last 200 years the extinction problem as it relates to modern species has not been considered seriously by scientists, much less by leaders of the business community, economists and politicians. As a result of the data and arguments that have been brought forth by researchers and conservationists recently, the effects of habitat destruction, invasive species, hunting, disease and associated factors have now been brought to the attention of the world. It can no longer be said that these issues are being ignored.

Setting the impossibility of preventing all extinction aside, thanks to the efforts of conservationists who understand how much humanity depends on biodiversity, and thanks to the efforts of researchers in many fields who are contributing to the improvement of our understanding of the processes that lead to extinction, people on all continents and in all walks of life are now aware of the problems posed by the unquestionably high rates of biodiversity loss the Earth is currently witnessing. More importantly, a greater proportion of people than at any time in humanity's past are aware of the need to conserve species as well as some of the solutions that have been developed to address this need. Research into these issues will continue. Our ability to assess the true extinction threat posed to individual species and to higher taxonomic groups will improve. But it is also becoming increasingly clear that humanity cannot afford to wait until the oftentimes slow progress of scientific research provides us with robust and definitive answers to the question of how much extinction risk is present in the world today. Whether the affects of human population growth and economic development are allowed to develop unchecked to the point at which the extinction of living species eventually does reach a magnitude consistent with the great extinctions in Earth history will depend on the decisions each individual makes with respect to how they live, what they buy, how they vote and what sort of planet they wish to pass on to future generations.

Reducing the current rate of species loss due to human influence will involve difficult decisions. But our species has an unparalleled capacity to seek out answers to difficult questions and to learn. On occasion human societies have decided collectively that some practices are so detrimental to the society's long-term survival these practices have come to be regarded as socially, commercially and morally unacceptable. Free speech, emancipation from slavery, gender equality, access to health care, access to education, care for the elderly and voting rights are all examples of decisions human cultures have made, not because they have been forced to or as a result of some short-term economic self-interest, but because these issues became so important to members of the collective human family they achieved the status of philosophical and moral imperatives. It can only be hoped that, over time, the responsibility each of us bears to preserve the diversity of our planet, not just for our own descendants but for the descendants of all the species with which we share this world, will achieve a similar moral status. The simple fact that public interest in extinction and in biodiversity-related issues has never been higher can only be interpreted as an encouraging sign.

Glossary

acritarch – organic-walled microfossil that cannot be placed in another taxonomic group, e.g. egg cases of small metazoans or resting cysts of many kinds of green algae.

actualism – the doctrine that the same (actual) processes observed to exist in the modern world existed in the past and can be invoked to account for or explain geological phenomena.

albedo – ability of a substance or surface to reflect radiation. A black surface has albedo of 0.0, a white surface an albedo of 1.0.

algae – very large group of simple autotrophic eukaryotes ranging from unicellular protists to multicellular marine plants, e.g. kelp.

ammonite – an extinct group of cephalopod molluscs closely related to modern coleoids (squid).

amphibian – a diverse group of ectothermic tetrapod vertebrates that lay non-aminote eggs, e.g. amphibians include frogs, salamanders, newts, and caceilians.

anagenesis – evolutionary change in which an entire population or set of populations is transformed to a new genetic state (also known as phyletic evolution or phyletic change).

anaerobe – an organism that is capable of living with as little as 0.1 ml dissolved oxygen per litre of water, e.g. anaerobic bacteria.

angiosperm – large and diverse group of seed producing plants including trees and shrubs which have enclosed seeds, and flowers and fruits. This group includes all flowering plants.

anoxic – an environment that contains no free oxygen, e.g. anaerobic marine bottom water.

Appalachia – the eastern portion of the North American continent including the piedmont area and the Appalachian mountains stretching from present day New Jersey to Alabama. This region remained above sea level in the Cretaceous during the time of the Western Interior Seaway.

archaeocyathid – primitive, marine, sessile, cup-shaped, reef building invertebrate organisms.

archosaur – a group of diapsid, amniote, tetrapod vertebrates that includes modern crocodiles and birds as well as all extinct dinosaurs and pterosaurs.

arthropod – a huge group of invertebrate organisms that possess an exoskeleton, segmented body, and jointed appendages including insects, arachnids, crustaceans, centipedes and scorpions.

Avalonia – a Palaeozoic island microcontinent that included parts of present day Great Britain, the east coast of North America, northern Scandinavia as well as parts of Belgium, France, Germany and Poland. Avalonia collided with Baltica and Laurentia during the Silurian and Devonian to form Euramerica which was eventually incorporated into Pangea during the Late Paleozoic. Subsequently Avalonia was rifted apart during the formation of the Atlantic Ocean basin.

background extinction – originally defined as extinctions that characterized those stratigraphic stages whose family-level magnitudes lay within two standard deviations of a linear regression of percent extinction on time. Conceptually this term refers to extinctions that result from the operation of normal competition and natural selection. Over 90% of all extinctions in earth history have taken place during times of background extinction.

Baltica – a Palaeozoic island microcontinent that included the eastern European craton of northwestern Eurasia. Baltica rode on a separate tectonic plate during the Early Palaeozoic, but collided with Avalonia and Laurentia during the Silurian and Devonian to form Euamerica which was eventually incorporated into Pangea during the Late Paleozoic. Subsequently Avalonia was rifted apart during the formation of the Atlantic Ocean basin.

belemnite – an extinct order of squid-like cephalopods that possess an internal bullet-shaped support structure (the guard).

benthic – the set of organisms which live on the bottoms of oceans, seas, and lakes.

biomass – the mass of living organic material in a given area or ecosystem.

biome – climatically or geographically defined regions of the Earth's surface or near subsurface that are characterized by similar environmental conditions and that support similar sorts of organisms.

biomere – a regional biostratigraphic unit bounded by abrupt non-evolutionary changes in the dominant elements of a single phylum or organisms. The significance of biomeres is that the upper biomere boundaries tend to coincide with the disappearance of all but a few trilobite (and other) species over very short stratigraphic interval.

biota – the total collection of organisms living in a geographic area, physiographic region or geological time interval.

biozone – the contiguous time and space-bounded region whose boundaries are defined by the occurrences of fossils.

bird – any feathered, winged, bipedal, endothermic (warm-blooded), egg-laying, amniotic vertebrate tetrapod.

bivalve – a large class of marine and freshwater molluscs characterized by a laterally compressed body enclosed in a hinged shell consisting of two distinct parts; including oysters, mussels, scallops, and clams.

bivalve (rudist) – a group of large marine bivalves with a cup-like morphology which constructed large reefs during the Late Cretaceous, especially in the Tethys Sea.

black shale – a deposit of fine-grained sedimentary rock rich in unoxidized organic matter. Usually regarded as evidence of low-oxygen or no-oxygen conditions.

blastoid – an extinct class of stalked echinoderm. Blastoids appear in the fossil record in the Ordovician.

blue-green algae – a phylum of bacteria (cyanobacteria) that obtain their energy through photosynthesis.

bolide – any extraterrestrial body, e.g. comet or meteor, in the 1–10 km size range. Bolide is used as a generic term for objects that have or might strike the Earth and leave a substantial crater as a mark of their impact.

brachiopod – a phylum of marine animals with a fleshy stalk or pedicle for attachment to the substrate, a laterally compressed body with a spiral feeding organ (lophophore) enclosed by pair of valves or shells. Brachiopods appear in the Early Cambrian and reach their maximum diversity in the Late Palaeozoic.

bryozoan – a phylum of aquatic solitary or colonial animals with a u-shaped gut and spiral circlet of tentacles (lophophore) surrounding a mouth. Bryozoans first appear in the Early Ordovician.

calcrete – a hardened crust of calcium carbonate that forms in arid and semiarid environments. Layers of calcrete in sedimentary deposits represent evidence for subareal exposure and arid evaporitic environmental conditions in the Earth's past.

Cambrian Evolutionary Fauna – assemblage of fossil groups accounting for a dominant proportion of the total fossil fauna in the Cambrian, including trilobites, inarticulate brachiopods, monoplacophoran molluscs, and hyoliths.

catastrophism – the theory that major physical and biotic changes in Earth's history have been precipitated through the action of sudden, short-term, violent global events that have no precedent in human history.

cephalopod – marine molluscs with strong bilateral symmetry, a prominent head with well-developed sense organs and a circlet of muscular tentacles surrounding a central mouth. Cephalopods first appear in the Late Cambrian and are extant today in the form of cuttlefish, squid, and *Nautilus*.

champsosaur – a member of the diapsid archosaur Order Choristodera; a group of semi-acquatic crocodile-like species that appear in the fossil record during the middle Jurassic and disappear from it in the Miocene.

Chicxulub crater – a large (*c*. 100 km wide), multi-ringed, circular morphological structure buried below Mexico's Yucatán Peninsula generally thought to be associated with the bolide impact at the K-Pg boundary though recent research has challenged this.

cladogenesis – a mode of evolutionary change in which a small, usually peripheral population or set of populations of a species is transformed via natural selection to a new genetic state such that a reproductive disjunction occurs between it and the parent species. This type of evolution is also known as 'branching evolution' and stands in contrast to anagenetic or phyletic evolution.

clathrate – a chemical compound with atoms that form a lattice or cage trapping the molecules of another substance in its structure.

Chondrichthyes – the group of non-amniotic vertebrate fish with an internal skeleton of cartilage rather than bone, including sharks, skates rays, and chimaeras. Chondrichthians first appear in the fossil record in the Ordovician.

confidence interval (stratigraphic) – a determination of the length of the stratigraphic interval (in terms of physical distance or time) above the last observed fossil occurrence of a species or other taxonomic group in which the existence of the taxon can be regarded as probable to a given level of certainty (e.g. 95%). Calculation of stratigraphic confidence intervals are based on the distribution of gaps between occurrences of the taxon during the known part of its stratigraphic range.

conifer – a group of cone-bearing gymnosperm plants consisting of trees and shrubs, including cedars, cypresses, firs, junipers, kauris, larches, pines, redwoods, spruces. Conifers appear in the Carboniferous and were the dominant large land plant from the Late Carboniferous through the Late Cretaceous.

conodont – an extinct group of pre-vertebrate chordates that resembled eels; usually represented by their tooth-like microfossils - elements of a complex feeding apparatus. Conodonts appear during the Early Cambrian and disappear by the Triassic-Jurassic boundary.

continental drift – the theory that the modern continental landmasses were once joined together into a single super-continent and subsequently drifted apart.

convergent plate boundary – in plate tectonics a region in which two tectonic plates are moving toward each other forcing the denser plate to be subducted beneath the lighter and causing deformation to occur in the internal structure of both plates.

coral – solitary or colonial multicellular marine animals characterized by a central mouth surrounded by a circlet of tentacles, a blind gut, and an three-layered body wall which is often surrounded by a had exoskeleton composed of calcium carbonate. Corals appear in the Cambrian Period and are reef framework builders throughout the Early Palaeozoic.

Coriolis Effect – the deflection of an object moving in a rotating reference frame when viewed by an observer located within the reference frame as a result of both the Coriolis and centrifugal forces acting on the moving object.

cosmopolitan species – a species whose geographic and environmental range is so broad that it includes most, if not all, relevant biomes; a globally distributed species.

crinoid – an extant class of stalked echinoderm. Crinoids appear in the Ordovician and reach their maximum diversity in the Carboniferous.

crustacean – a very large group of arthropods characterized by a tripartite subdivision of the segmented body (head, thorax, abdomen). Crustaceans appear in the middle Cambrian and their numbers and ecological roles increase from that time to now.

cycad – a group of gymnosperm seed plants with a stout and woody (ligneous) trunk and a crown of large, hard and stiff, evergreen leaves. Cycads appear during the Early Permian.

death assemblage – an association of organisms or fossils that had been brought together after their deaths, usually by the processes of erosion and redeposition.

Deccan Traps – a large accumulation of igneous and volcanic rock consisting mostly of successive layers of erupted flood basalts that was emplaced on the Indian subcontinent at the end of the Cretaceous and beginning of the Paleogene, between 60 and 68 million years ago.

deep-sea trench – long, narrow depression of the sea floor that marks the front along which one oceanic tectonic plate is being subducted beneath another.

diatom – a large group of solitary or colonial unicellular phytoplankton characterized by their distinctive siliceous endoskeletons or frustules. Diatom frustrules appear in the latest Triassic.

dinoflagellate – a large group of aquatic (mostly marine), solitary or colonial unicellular flagellated phytoplankton. Dinoflagellate cysts appear in the Late Triassic.

dinosaur – a clade of terrestrial, archosaurian tetrapods including traditional dinosaurs (e.g. *Tyrannosaurus*, *Triceratops*) and their lineal descendants, birds. Dinosaur fossils appear during the Early Triassic and all non-avian dinosaur species vanish from the fossil record by the end of the uppermost Cretaceous (Maastrichtian Age). (See non-avian dinosaur.)

divergent plate boundary – in plate tectonics a region in which two tectonic plates are moving in opposite directions across a spreading centre (e.g. mid-oceanic ridge, volcanic rift) at which new oceanic crust is being created via volcanism.

Drake Passage – body of water between the southern tip of Chile (Cape of Good Hope) and the South Shetland Islands. This passage was closed until *c.* 41 million years ago at which time the northward drift of South America allowed a deep ocean basin to form between these two continents.

dropstone – isolated rocks or rock fragments found in fine-grained sediments or sedimentary rocks that can range in size from pebbles to boulders and are usually associated with a lack of grain size grading that would indicate movement by normal sediment transport processes. These rocks appear to have been dropped vertically through the water column as single, isolated events. Dropstones provide evidence of ancient glaciers, ice caps, sea ice, and icebergs.

dysaerobic – an aquatic environment with between 0.1 and 1.0 ml of dissolved oxygen per litre of water.

echinoid – a member of a large group of globular to disk-shaped, free-living, marine echinoderms also known a sea urchins. Regular echinoids (pentameral + radial symmetry) are predators that first appear in the fossil record during the Ordovician Period. Irregular echinoids (pentameral + bilateral symmetry) tend to be filter or deposit feeders that first appear in the Jurassic Period, shortly after the appearance of the modern phytoplankton groups.

echinoderm – a member of a large phylum-level clade of invertebrate, marine, metazoans including cystoids, blastoids, crinoids, asteroids, ophiuroids, and echinoids. All species within this group have an initial bilateral symmetry that changes to pentameral symmetry as the individual develops. Echinoderms appear in the Early Cambrian.

epicontinental sea – a large shallow sea formed as a result of sea level rising to flood the interiors of continental platforms. These seas often supported rich assemblages of marine organisms, but were unstable ecologically in that a local, regional, or global sea-level fall could drain the continental interior over a relatively brief time interval leading to extinction of the indigenous biota.

Euamerica – a continent formed in the Devonian as the result of continent-continent collisions between Avalonia, Baltica, and Laurentia. Euamerica was a island continent through the upper Palaeozoic and eventually became incorporated into Pangea during the Permian.

eurypterid – an extinct group of aquatic arachnid arthropods.

eustatic sea-level change – sea-level changes that involve variation in the global level of sea water.

evaporite – a deposit of mineral crystals or sequence of mineral species that result from the evaporation of natural waters, especially sea water. Presence of evaporite deposits constitutes geological evidence of arid environmental conditions.

Ferrel Cell – a zone of open circulation in the atmosphere in the mid-latitudes between the Hadley and Polar cells that is the result of eddy circulation driven by the latter two closed cells.

foraminifera (planktonic, benthic, nummulitid, fusulinid, textularid) – a large group of amoeboid protists with granular cytoplasm, reticulate pseudopods, and (in most species) a single or multi-chambered shell or test composed of calcium carbonate or agglutinated sediment particles.

fusuilinacean – member of a superfamily of large benthic foraminiferans that secret complex multi-chambered ovoid to cigar-shaped shells or tests. Fusulinid fossils appear in the Devonian and the group becomes extinct at the end of the Permian.

gastropod – member of a large and diverse class of aquatic and/or territorial mollusc having (in most species) a univalved shell and a broad foot used for locomotion including snails, slugs and nudibranchs.

ginko – a small group of large gymnosperm trees with unusual morphological attributes (e.g. seeds not protected by the ovary wall). \ Ginkos do not appear to be closely related to any living tree species.

Gondwana – the southernmost of two large Late Palaeozoic continents that tectonically fused to form Pangea in the Permian. Gondwana was composed of the modern continents of Antarctica, Australia, Africa, India, and South America along with the subcontinents of Madagascar and the Arabian Peninsula.

graptolite – members of an extinct class of morphologically distinctive colonial planktonic hemichordates that appear in the upper Cambrian Period and become extinct by the lower Carboniferous after a long period of declining species numbers.

greenhouse conditions – intervals in Earth history characterized by warm-arid conditions, few glaciers, high sea-levels, and high average sea-surface temperatures with relatively little variation between polar and tropical regions.

greenhouse gas – a gas that emits absorbed radiation (e.g. sunlight) within the thermal infrared range. The primary naturally occurring atmospheric greenhouse gases are water vapour, carbon dioxide (CO_2), methane (CH_4), nitrous oxide (N_2O), and ozone (O_3).

gymnosperm – a large and diverse group of seed-producing trees and shrubs having unenclosed seeds including conifers, cycads, and ginkos. Gymnosperm fossils first appear in the Late Carboniferous.

gypsum – a soft, sulphate mineral ($CaSO_4 \cdot 2H_2O$) that typically forms as an alteration product of anhydrite in an evaporite sequence.

Hadley Cell – a well defined zone of closed atmospheric circulation characterized by rising warm air near the equator, poleward flow at altitudes of 10-15 km, descending cool air in the subtropics (*c.* 30˚N-S latitude), and equatorward flow at the surface. In addition to this the Coriolis Effect imparts an clockwise lateral circulation pattern to the northern Hadley Cell and a counterclockwise lateral circulation to the southern Hadley Cell.

halkieriid – an extinct mollusc-like benthic marine invertebrate animal of uncertain phylogenetic affinity whose long, relatively flat, bilaterally symmetrical body was covered by calcareous scales or sclerties with two large sclerties — the shell plates — located along the midline at the dorsal and ventral ends. Halkerid fossils are found only in sediments of lower and middle Cambrian age.

hiatus (stratigraphy) – a gap or discontinuity in the temporal sequence of sediments in a stratigraphic succession. Hiatuses can be produced by non-deposition or by active erosion of sediments deposited previously.

hyolith – an extinct mollusc-like benthic marine invertebrate animal of uncertain phylogenetic affinity that constructed a small conical operculate shell with two laterally directed ventral support structures (the helens). Hyoliths appear at the base of the Cambrian and become extinct in the Permian.

icehouse conditions – intervals in Earth history characterized by cool-wet conditions, extensive glaciation, low sea-levels, and cool average sea-surface temperatures with relatively pronounced variation between polar and tropical regions.

ichthyosaur – a relatively diverse group of large-sized, marine, amniote, tetrapods (marine reptiles) with highly modified dolphin-like bodies adapted for high swimming speeds in open water. Ichthyosaurs appear in the Early Triassic and become extinct by the middle Cretaceous.

index fossil – a fossil species or group that is easily recognizable, exhibits a wide geographic range with short dispersal times, occurs in many different environments, is characterized by a high rate of evolutionary change, and became extinct rapidly such that its first and last appearances in distant stratigraphic successions denote approximately time-synchronous horizons. Index fossils are used for inferring physical and temporal relations between packages of sedimentary rock strata.

insect – members of a large class of arthropods characterized by a chitinous exoskeleton, three primary body segments (head, thorax, and abdomen), three pairs of jointed legs, compound eyes, and one pair of antennae. Insects first appear in the fossil record in the mid-Devonian.

iridium – a rare-earth chemical element of the platinum group that, in its pure form, is very a hard, brittle, silvery-white metal. Iridium is abundant in meteorites (and in the Earth's core). Its presence in stratigraphic successions is usually attributed to volcanogenic or extraterrestrial input.

isotope – variants of a chemical element that have the same number of protons, but different numbers of neutrons causing the mass of the element to change.

Larimidia – the western island continent that was created when the interior of the North American tectonic plate was flooded by a sea-level high-stand (producing the Western Interior Seaway) during the Cretaceous.

large igneous province (LIP) volcanism – an extremely large (> 100,000 km²) emplacement of intrusive and/or extrusive igneous rock over a geologically short time interval (= a few million years).

Laurentia – the ancient geological core of the North American continent that rides atop the North American tectonic plate.

life assemblage – an association of organisms that existed when the organisms were alive that has been preserved in the fossil record.

lycopods – a group of trachaeophyte vascular plants that possess microphyllous leaves that have only one vein.

lystrosaur – a genus of Late Palaeozoic dicynodont therapsid that occurred in Gondwana (Antarctica, India, South Africa. Lystrosaur appear in the Late Permian and survive the end-Permian extinction, but become extinct in the Early Triassic.

mammal – any air breathing, endothermic (warm-blooded), amniote, vertebrate tetrapod with three middle ear bones and mammary glands that are functional in mothers with young. The first mammal fossils appear in the Late Triassic.

mammoth – any member of the extinct probiscidean genus *Mammuthus* which is characterized by long, curving tusks and a body covering of long hair.

mass balance calculation – an application of the physical principle of the conservation of mass in which the magnitudes to input factors (e.g. elements, molecules, isotopes) are estimated in such a way that either measured or anticipated system outputs (e.g. global temperature rise/fall) are accounted for (subject to certain assumptions).

mass extinction – a term used to denote an unusually large extinction event as recorded in the fossil record, but formally defined by as the extinctions that characterized those stratigraphic stages whose family-level magnitudes lay beyond two standard deviations of a linear regression of percent extinction on time. Conceptually mass extinctions are often regarded as those that result from the operation of processes other than normal competition and natural selection. Less than 10% of all extinctions in Earth history have taken place during times of mass extinction.

mastodon – any member of the extinct probiscidean genus *Mammut* which is characterized by molar teeth with low conical projections on the occlusal surface.

megafauna – collective term for animal of large adult body size (e.g. >44 kg [100 lbs]).

mid-ocean ridge – a general term for an underwater mountain system that forms around linear zones of weakness in the oceanic crust beneath the upwelling limbs of mantle convection currents at which new oceanic crust is emplaced as a series of igneous intrusions and extruded volcanics. A mid-ocean ridge demarcates the boundary between two tectonic plates at divergent plate boundaries.

Milankovitch cycles – patterns of cyclic variation in the Earth's climate caused by changes in the Earth's orbtial parameters, including orbital shape (eccentricity, period 413,000 years), tilt of the planets rotations axes (obliquity, period 41,000 years), spatial orientation of the rotational axes relative to a fixed reference (axial precession, period: 26,000 years), orientation of the orbital ellipse relative to a fixed reference (apsidal precession, period: 21,000-25,000 years), and orbital inclination relative to the mean plane of the solar system (orbital inclination, period: 100,000 years).

Modern Evolutionary Fauna – assemblage of fossil groups that account for a dominant proportion of the total fossil fauna in the interval from the Early Triassic to the Recent, including demosponges, gymolamatid bryozoans, gastropod and bivalve molluscs, echinoid echinoderms, sharks and rays and bony fish.

mollusc – a member of a large and diverse animal phylum whose species are characterized by common possession of a mantle (= dorsal part of the body wall used for gas exchange, excretion, and is the site of shell/spine/sclerite secretion) and details of the organization of the nervous system.

mosasaur – a moderately diverse group of large, semi-aquatic and aquatic varanoid lizards. Following the extinction of the ichthyosaurs and plesiosaurs/pliosaurs mosasaurs were the dominant Late Cretaceous marine predator. Mosasaurs appear in the Turonian Stage of the Late Cretaceous and become extinct in the Maastrichtian Stage, though no mosasaur fossils have been found at the top of the Maastrichtian (near the K-Pg, boundary).

multiple interacting cause (MIC) scenario – the class of mass extinction cause hypotheses that invokes interactions between multiple mechanisms (e.g. large igneous province eruption, seal-level fluctuation, bolide impact) that coincidently occur over the same interval of Earth history as being responsible for some or all historical peaks in extinction intensity.

nanoplankton – plankton of small size c. 2–20 μm in the longest dimension.

natural selection – the natural process that sorts individuals in terms of their ability to survive environmental change and or win competitive interactions, and so make a differentially large contribution to succeeding generations as the result of the possession of heritable variations in morphology, physiology, behaviour, etc.

non-avian dinosaur – a collective term used to refer to all members of the clade Dinosauria except birds; the group of traditional or classic dinosaurs.

organelle – a separate functional unit within a cell that has a specific function or set of functions and is often enclosed in its own lipid layer (= cell wall). Many organelles are thought to have originated as separate organisms whose activities were incorporated into the body of a composite organism that was the ancestor of the eucaryotic cell.

ostracod – members of a typically small sized group of crustaceans characterized by lateral flattening of the body which is enclosed by a hinged pair of shells of valves composed of chitin or calcium carbonate.

overkill hypothesis – the scientific scenario that accounts for the Pliocene-Pleistocene megafauna extinctions in North America, South America, and Eurasia as the result of human populations that were new migrants into these areas at the time of the extinctions.

ozone – a naturally occurring molecule consisting of three oxygen atoms (O_3).

Palaeozoic Evolutionary Fauna – an assemblage of fossil groups that account for a dominant proportion of the total fossil fauna in the interval from the Early Ordovician to the Late Permian, including anthrozoan cnidarians, stenolaematid bryozoans, cephalopod molluscs, articulate brachiopods, ostracod arthropods, and crinoid and stelleroid echinoderms.

Pangea – the Late Palaeozoic supercontinent that formed as a result of the tectonic collision of Euamerica and Gondwana.

Panthalassa – the super-ocean that was formed as a result of Pangea's assembly and was later subdivided into Atlantic and Pacific regions as a result of Late Triassic rifting that led to formation of the Atlantic Ocean Basin.

palaeosol – a preserved former soil deposit incorporated into the stratigraphic record by burial beneath other types of sediments (e.g. marine sands and muds that form as a result of eustatic sea-level rise.)

paraequatorial region – a physiographic region of warm equable environmental conditions in the area immediately adjacent to the equatorial zone.

paratropical – a physiographic region of warm equable environmental conditions in the area immediately adjacent to the tropical zone.

physiology – the study of how organisms, organ systems, organs, cells, and bio-molecules carry out the chemical or physical functions necessary to maintain living systems.

phytoplankton – the component of the plankton community composed of organisms that can produce complex organic molecules from simpler material by using energy supplied from the ambient environment (e.g. sunlight, thermal energy).

placoderm – an extinct class of prehistoric fish characterized by articulated armoured plates covering the head and thorax region.

placodont – a group of extinct air breathing, Triassic sauropterygian amniote tetrapods whose teeth were modified for crushing the shells of shallow-water benthic marine invertebrate species (e.g. gastropods, bivalves). Placodonts appear in the mid-Triassic and become extinction in the Late Triassic.

plankton – the group of organisms that live suspended in the water column and are unable to control their lateral position in opposition to prevailing currents.

plate tectonics – the mid-twentieth century update of the theory of continental drift that provided empirical evidence for the mechanism of sea-floor spreading.

plesiosaur – a member of a group of large-sized, sauropterygian-grade, air breathing, amniote tetrapods with a broad body, short tail and two sets of limbs that have been modified into paddles for power and steering control. Plesiosaurs appear in the Early Jurassic and continue to their extinction in the Late Cretaceous (Maastrichtian Stage).

Polar Cell – a well-defined zone of closed atmospheric circulation characterized by rising relatively warm air near the 60˚ N-S latitudes, poleward flow at altitudes of c. 8 km, descending cool air in the polar region, and equator-ward flow at the surface. In addition to this the Coriolis Effect imparts an clockwise lateral circulation pattern to the northern Polar Cell and a counterclockwise lateral circulation to the southern Polar Cell.

population – an interacting group of individuals of the same species that inhabit the same physical area or region.

Principle of Superposition – the observation that sedimentary layers at the bottom of a stratigraphic sequence must be older than those at the top because the bottom layers must have already existed in order for the upper layers to have been deposited on top of them. Superposition is a key tool for inferring the relative sequence of events in earth history.

pterosaur – a member of a group of extinct, winged, archosaur-grade air-breathing, amniote tetrapods whose phylogenetic relations are (presently) uncertain, but which are characterized by a large number of adaptations that enabled individuals to engage in powered flight. The first pterosaur fossils appear in the Late Triassic and the last from the latest Cretaceous Maastrichtian stage, though not from sediments coincident with the K-Pg boundary.

Pull of the Recent – a systematic bias in biodiversity estimates of younger (as opposed to older) stratigraphic intervals engendered by the fact the many species, genera, and higher taxonomic groups have a spotty occurrence in the fossil record but are represented in the modern biota by numerous extant representatives. This results in stratigraphic ranges for the groups present in the modern biota being extended to the Recent and being counted as occurring in all stratigraphic intervals between their first appearance in the fossil record and the Recent.

quartz (shocked) – a form of quartz in which the crystal structure has been deformed at the microscopic level - creating linear planar features - by intense pressure and limited temperature increase for a short duration.

Shocked quartz can be produced by nuclear explosions and by the impact of a bolide-sized extraterrestrial object.

radiation (ultraviolet) – radiation whose wavelength is 10–400 nm. Although ultraviolet (UV) radiation is invisible to human eyes it carries high energies and can damage unprotected tissues. Ozone in the Earth's atmosphere acts as an UV-radiation filter, by absorbing its energy and emitting longer wavelength infrared radiation.

radioisotope – atoms that have unstable atomic configurations and, over time, decay into simpler substances by emitting radiation and subdividing themselves into lighter daughter atoms. This information can be used to estimate the ages of various materials.

radiolarian – a diverse, extant group of amoeboid protists characterized by an internal skeleton composed of opaline silica, stiffened axopodial pseudopodia and a nuclear region that is separated from the cytoplasm in the remainder of the call by a porous organic-walled central capsule. The first radiolarian fossils have been recovered from the Early Cambrian.

Red Queen Hypothesis – the idea that within any system undergoing evolution by natural selection any individual species' or groups' state of adaptation is always provisional and subject to change depending on evolutionary developments in other local competitor or prey species; the principle that continuing adaptation is needed in order for a species to maintain its relative fitness among the local species with which it is co-evolving.

regression (sea level) – any consistent and sustained reduction in the level of the oceans that takes place at local, regional, or global scales.

reptile – a class of air-breathing, egg-laying, amniote tetrapods that includes turtles, lizards, snakes and crocodiles. Reptile fossils first appear in the Carboniferous Period.

single cause (SIC) scenario – the class of mass extinction cause hypotheses that invokes a single mechanism (usually bolide impact) as being responsible for some or all historical peaks in extinction intensity.

shale – a fine-grained, sedimentary rock composed of clay minerals and tiny fragments of other minerals.

speciation – the creation of a new species by one of a diverse set of of natural processes.

species-area effect – a regular mathematical relation between the number of species inhabiting an area or region and the size of the area or region.

stable isotope – an isotope of an atom that is not subject to spontaneous decay under normal environmental conditions.

stromatolite – layered accretionary structures formed in shallow water by the trapping, binding and cementation of sedimentary grains by mats of microorganisms (e.g. cyanobacteria); stromatolites provide one of the most ancient lines of evidence for the antiquity of life on Earth.

stromatoporoid – a class of aquatic sclerosponges that were important Palaeozoic reef framework builders. The first stromatoporoid fossils appear in the Ordovician and their extinction occurs in the Cretaceous after a long period of taxonomic diversity decline.

subareal – a phenomenon or process that takes place on the Earth's surface; beneath the atmosphere.

suture zone – the line of collision between two tectonic plates at a convergent plate boundary, often marked by a mountain range created by the deformation of rocks and sediments along the tectonic front as a result of the pressures of the colliding plates.

Tasmanian Gateway – the seaway that exists between Australia and Antarctica which opened to deep-water circulation as a result of plate tectonic sea-floor spreading during the Eocene-Oligocene transition.

taxa/taxon – group of populations all of which are judged to be composed of the same species, genus, family or other taxonomic group.

tektite – small bodies of typically black, but also green, brown or grey, natural glass characterized by a fairly homogeneous composition and low water content. For the most part tektites are regarded as terrestrial material forcibly melted and ejected from an impact crater that has solidified in the atmosphere and fallen back to Earth as a part of a tektite strew-field.

terror bird – members of the Family Phorusrhacidae, a clade of large carnivorous flightless birds that were the dominant terrestrial predators in South America when it was an island continent, but which were largely eliminated through competition with placental mammals during the Great American Faunal Interchange.

Tethys Sea – the broad, but shallow ocean that existed off the east coast of Pangea to the south of Laurasia and east of Gondwana. Through the Mesozoic and Caenozoic the Tethys Sea became progressively smaller and more restricted. It exists in remnant form today as the Mediterranean and Black seas.

thecodont – a group of archosaurian-grade, air-breathing, amniote tetrapods with teeth set in sockets within both the cranium and mandible.

therapsid – a group of advanced synapsid tetrapods with hair, lactation and an erect posture. The earliest mammals are thought to have evolved from this group. Therapsid fossils first appear in the mid-Permian.

transgression (sea level) – any consistent and sustained increase in the level of the oceans that takes place at local, regional, or global scales.

trilobite – a large group of extinct, marine arthropods characterized by the presence of an exoskeleton composed of calcium carbonate and subdivided anterio-posteriorly into a cephalon (head), abdomen, and pygidium (tail). Trilobite fossils first appear at the base of the Cambrian and make their last (extinction) appearance in the Late Permian.

topical zone – the geographic zone centred at the equator and extending *c.* 23° N-S latitude which is typically characterized by warm, humid environmental conditions.

uniformitarianism – in its original form this term refers to the principle that Earth history constitutes a series of infinitely repeating cycles of uplift, deformation, erosion, and sedimentation. Later it was reformulated to include only four uniformities that unite Earth's present with its geological past: law, methodology, kind and degree.

Western Interior Seaway – a broad, but relatively shallow sea that existed in the middle section of North America throughout most of the Cretaceous owing to a pronounced sea-level highstand.

whales – any member of the mammalian Order Cetacea which are large, air-breathing, placental mammals with a fusiform body shapes, laterally directed tail flukes, and a dorsally directed blowhole, Whales appear in the early Eocene in a semi-aquatic state, and with the first species to achieve an obligate aquatic lifestyle in the mid-Miocene to Pliocene interval.

Index

References

INTRODUCTION

Alvarez, W., 1997. *T. rex and the Crater of Doom*. Princeton University Press, Princeton.

Alvarez, L.W., *et al.*, 1998. Extraterrestrial cause for the Cretaceous–Tertiary extinction. *Science*, 208: 1095–1108.

Archibald, J.D., *et al.*, 2010. Cretaceous extinctions: multiple causes. *Science*, 238: 973.

Schulte, P., *et al.*, 2010. The Chicxulub Asteroid impact and mass extinction at the Cretaceous–Paleogene boundary. *Science*, 327: 1215–1218.

Ward, P.D., 1995. *The End of Evolution: Dinosaurs, Mass Extinction and Biodiversity*. Weidenfeld and Nicolson, London.

CHAPTER 1

Van Valen, L., 1973. A new evolutionary law. *Evolutionary Theory*, 1: 1–30.

CHAPTER 2

Darwin, C., 1859. *On the Origin of Species by Means of Natural Selection, or the Preservation of Favoured Races in the Struggle for Life*. John Murray, London,

Foote, M. and Raup, D.M., 1996. Fossil preservation and the stratigraphic ranges of taxa. *Paleobiology*, 22 (2): 121–140.

Newell, N.D., 1963. Crises in the history of life. *Scientific American*, 208: 76–92.

Newell, N.D., 1967. Revolutions in the history of life. *In:* Albritton, C.C., *et al.* (eds). *Uniformity and Simplicity: A Symposium on the Principle of the Uniformity of Nature*. Geological Society of America Special Paper 89, Boulder, CO, pp. 63–91.

Raup, D.M., 1991. *Extinction: Bad Genes or Bad Luck*. W.W. Norton and Co., New York.

Schindewolf, O. H., 1954. Über die möglichen Uraschen der grossen erdgeschictlichen Faunenschnitte. *Neues Jarbuch für Geologie und*

Paläontologie Monatshefte, 1954: 457–465.

Simpson, G.G., 1944. *Tempo and Mode in Evolution*. Columbia University Press, New York.

Smith, A.B. and McGowan, A.J., 2007. The shape of the Phanerozoic marine palaeodiversity curve: how much can be predicted from the sedimentary rock record of western Europe. *Palaeontology*, 50: 765–774.

CHAPTER 3

Benton, M.J., 1993. *The Fossil Record 2*. Chapman & Hall, London, 845 pp.

Harland, W.B., *et al.*, 1967. *The Fossil Record: A Symposium with Documentation*. Geological Society of London Special Publication, no. 2, 827 pp.

Oreskes, N., 1998. The rejection of continental drift. *Historical Studies in the Physical and Biological Sciences*, 8(1): 1-39.

Oreskes, N., 1999. *The Rejection of Continental Drift: Theory and Method in American Earth Science*. Oxford University Press, Oxford.

Sepkoski, J. J., Jr. 1997. Biodiversity: past, present, future. *Journal of Paleontology*, 71:533–539.

CHAPTER 4

Gould, S.J., 1987. *Time's Arrow, Time's Cycle: Myth and Metaphor in the Discovery of Geological Time*. Harvard University Press, Cambridge, MA.

Harland, W.B., *et al.*, 1967. *The Fossil Record: A Symposium with Documentation*. Geological Society of London Special Publication, no. 2, 827 pp.

Foote, M., 1994. Temporal variation in extinction rick and temporal scaling of extinction metrics. *Paleobiology*, 20(4): 424–444.

Jablonski, D., 1986. Background and mass extinctions: the alteration of macroevolutionary regimes. *Science*, 231: 129–133.

Jablonski, D., 1995. Extinctions in the fossil record. *In:* Lawton, J.H. and May, R.M. (eds), *Extinction Rates*. Oxford University Press, Oxford, pp. 25–44.

MacLeod, N., 2003. The causes of Phanerozoic extinctions. *In:* Rothschild, L. and Lister, A. (eds), *Evolution on Planet Earth*. Academic Press, London, pp. 253–277.

MacLeod, N., 2004. Identifying Phanerozoic extinction controls: statistical considerations and preliminary results. *In:* Beaudoin, A.B. and Head, M.J. (eds), *The Palynology and Micropaleontology of Boundaries*. Special Publications Volume 230, Geological Society of London, London, pp. 11–33.

Martin, R.E., 1996. Secular increase in nutrient levels through the Phanerozoic: implications for productivity, biomass, and diversity of the marine biosphere. *Palaios*, 11: 209–219.

Newell, N.D., 1963. Crises in the history of life. *Scientific American*, 208: 76–92.

Newell, N.D., 1967. Revolutions in the history of life. *In:* Albritton, C.C., *et al.* (eds), *Uniformity and Simplicity: A Symposium on the Principle of the Uniformity of Nature*. Geological Society of America Special Paper 89, Boulder, CO, pp. 63–91.

Phillips, J., 1860. *Life on Earth: Its Origin and Succession*. MacMillan, Cambridge.

Raup, D.M., 1991. *Extinction: Bad Genes or Bad Luck*. W.W. Norton and Co., New York.

Raup, D.M. and Sepkoski, J.J., Jr., 1982. Mass extinctions in the marine fossil record. *Science*, 215: 1501–1503.

Raup, D.M. and Sepkoski, J.J., Jr., 1986. Periodic extinction of families and genera. *Science*, 231, 833–836.

Rudwick, M.J.S. and Cuvier, G., 1997. *Georges Cuvier, Fossil Bones, and Geological Catastrophes: New Translations and Interpretations of the Primary Texts*. University of Chicago Press, Chicago.

Sepkoski, J.J., Jr., 1981. A factor analytic

description of the Phanerozoic marine fossil record. *Paleobiology*, 7(1): 36–53.

Sepkoski, J.J., Jr., 1982. *A Compendium of Fossil Marine Families*. Milwaukee Public Museum Contributions in Biology and Geology 51, 125 pp.

Sepkoski, J. J., Jr., 2002. A Compendium of Fossil Marine Animals. *In:* Jablonski, D., and Foote, M., (eds), *Bulletins of American Palaeontology* 363: Palaeontological Research Institution, Ithaca, New York, pp.1–563.

Van Valen, L., 1973. A new evolutionary law. *Evolutionary Theory*, 1: 1–30.

CHAPTER 5

Benton, M.J., 1990. Scientific methodologies in collision: the history of the study of the extinction of dinosaurs. *Evolutionary Biology*, 24: 371–400.

Hallam, A., 1992. *Phanerozoic Sea-Level Changes*. Columbia University Press, New York.

Haq, B., 1991. Sequence stratigraphy, sea-level change and significance for the deep sea. *International Association of Sedimentologists, Special Publication*, 12: 3–39.

CHAPTER 6

Landing, E., and MacGabhann, B. A., 2010. First evidence for Cambrian Glaciation Provided by Sections in Avalonian New Brunswick and Ireland: Additional data for Avalon–Gondwana Separation by the Earliest Palaeozoic. *Palaeoceanography, Palaeoclimatology, Palaeogeography*, 285(3–4): 174–185.

Palmer, A.R., 1965. Biomere: a new kind of biostratigraphic unit. *Journal of Paleontology*, 39: 149–153.

Palmer, A.R., 1984. The biomere problem: evolution of an idea. *Journal of Paleontology*, 58: 599–611.

Seilacher, A., 1984. Late Precambrian and Early Cambrian Metazoa: preservational or real extinctions? *In:* Holland, H.D. and Trendall, A.F. (eds), *Patterns of Change in Earth Evolution*. Springer-Verlag, Heidelberg, pp. 159–168.

Seilacher, A., 1989. Vendozoa: organismic construction in the Proterozoic biosphere. *Lethaia*, 17: 229–239.

CHAPTER 7

Sheehan, P.M., 2001. The Late Ordovician mass extinction. *Annual Review of Earth and Planetary Science*, 29: 331–364.

Stanley, S.M., 1987. *Extinctions*. W.H. Freeman and Co., New York.

CHAPTER 8

Buggisch, W., 1991. The global Frasnian-Famennian 'Kellwasser event'. *Geologische Rundschau*, 80:49–72.

Keller, G., 2005. Impacts, volcanism and mass extinction: random coincidence or cause and effect? *Australian Journal of Earth Sciences*, 52: 725–757.

CHAPTER 9

Becker, L., Poreda, R. J., Basu, A. R., Pope, K.O., Harrison, T. M., Nicholson, C. and Lasky, R., 2004. Bedout: a possible end-Permian impact crater offshore of northwestern Australia. *Science*, 304(5676):1469–1476.

McGhee, G.R., Jr., 1996. *The Late Devonian Mass Extinction: the Frasnian/Famennian Crisis*. Columbia University Press, New York.

Retallack, G. J., Seyedolali, A., Krull, E. S., Holser, W. T., Ambers, C. P. and Kyte. F. T., 1998. Search for evidence of impact at the Permian-Triassic boundary in Antarctica and Australia. *Geology*, 26:979–982.

CHAPTER 10

Benton, M.J., 2003. *When Life Nearly Died: the Greatest Mass Extinction of All Time*. Thames & Hudson, London.

Tanner, L.H., Lucas, S.G. and Chapman, M.G., 2004. Assessing the record and causes of Late Triassic extinctions. *Earth-Science Reviews*, 65:103–139.

CHAPTER 11

Archibald, J.D., 2011. *Extinction and Radiation: How the Fall of Dinosaurs Led to the Rise of Mammals*. Johns Hopkins University Press, Baltimore, MD.

Canudo, J. I., Keller, G. and Molina, E., 1991. Cretaceous/Tertiary boundary extinction pattern and faunal turnover at Agost and Caravaca, S. E. Spain. *Marine Micropaleontology*, 17:319–341.

Chenet, A.-L., *et al.*, 2009. Determination of rapid Deccan eruptions across the Cretaceous–Tertiary boundary using paleomagnetic secular variation: 2. Constraints from analysis of eight new sections and synthesis for a 3500-m-thick composite section. *Journal of Geophysical Research*, 114: BP6013–BP6038.

Courtillot, V., 1999. *Evolutionary Catastrophes: the Science of Mass Extinction*. Cambridge University Press, Cambridge.

Keller, G., *et al.*, 2004. Chicxulub impact predates the K–T boundary mass extinction. *Proceedings of the National Academy of Sciences*, 101: 3753–3758.

Keller, G., *et al.*, 2011. Deccan volcanism linked to the Cretaceous–Tertiary boundary mass extinction: new evidence from Ongc wells in the Krishna–Godavari Basin. *Journal of the Geological Society of India*, 79: 399–428.

MacLeod, N., 2004. Identifying Phanerozoic extinction controls: statistical considerations and preliminary results. *In:* Beaudoin, A.B. and Head, M.J. (eds), *The Palynology and Micropaleontology of Boundaries*. Special Publications Volume 230, Geological Society of London, London, pp. 11–33.

MacLeod, N., *et al.*, 1997. The Cretaceous–Tertiary biotic transition. *The Journal of the Geological Society of London*, 154: 265–292.

Smit, J., and ten Kate, W. G. H. Z., 1982. Trace element patterns at the Cretaceous-Tertiary boundary - consequence of a large impact. *Cretaceous Research*, 3:307–332.

CHAPTER 12

MacLeod, N., 2005. Mass extinction causality: statistical assessment of multiple-cause scenarios. *Russian Journal of Geology and Geophysics*, 9:979–987.

Prothero, D.R., 1994. *The Eocene–Oligocene Transition: Paradise Lost*. Columbia University Press, New York.

Prothero, D.R. and Berggren, W.A., 1992. *Eocene–Oligocene Climatic and Biotic Evolution*. Princeton University Press, Princeton, NJ, 568 pp.

Zachos, J.C., *et al.*, 2005, Rapid acidification of the ocean during the Paleocene-Eocene thermal maximum. *Science*, 2005, 308:1611–1615.

CHAPTER 13

Barnosky, A.D., *et al.*, 2004. Assessing the causes of Late Pleistocene extinctions on the continents. *Science*, 306: 70–75.

Haynes, G., 2009. *American Megafaunal Extinctions at the End of the Pleistocene.* Springer, New York.

Raup, D.M. and Sepkoski, J.J., Jr., 1984. Periodicity of extinctions in the geologic past. *Proceedings of the National Academy of Sciences of the United States of America*, 81: 801–805.

Raup, D.M. and Sepkoski, J.J., Jr., 1986. Periodic extinction of families and genera. *Science*, 231: 833–836.

CHAPTER 14

Leakey, R. and Lewin, R., 1996. *The Sixth Extinction*. Weidenfield, London.

Steadman, D.W., 1995. Prehistoric extinctions of Pacific island birds: biodiversity meets zooarchaeology. *Science*, 267: 1123–1131.

Vermeij, G.J., 2004. Ecological avalanches and the two kinds of extinction. *Evolutionary Ecology Research*, 6: 315–337.

CHAPTER 15

Courtillot, V., Jaeger, J.-J., Yang, Z., Féraud, G. and Hofmann, C., 1996. The influence of continental flood basalts on mass extinctions: where do we stand? Pp. 513–525. *In:* Ryder, G., Fastovsky, D. and Gartner, S., (eds.). *The Cretaceous-Tertiary event and other catastrophes in earth history*. The Geological Society of America, Special Paper, 307.

Grieve, R., Rupert, J., Smith, J. and Therriault, A.,1996. The record of terrestrial impact cratering. *GSA Today* 5:193–195.

Hallam, A., 1992. *Phanerozoic Sea-level Changes*. Columbia University Press, New York.

MacLeod, N., 2004. Identifying Phanerozoic extinction controls: statistical considerations and preliminary results. *In* Beaudoin, A.B., and Head, M.J., (eds.), *The palynology and micropaleontology of boundaries*, Special Publications 230, Geological Society of London, London, pp. 11–33.

MacLeod, N., 2005. Mass extinction causality: statistical assessment of multiple-cause scenarios: *Russian Journal of Geology and Geophysics*, 9:979–987.

Picture credits

Acknowledgements

There are many people I need to recognize and thank for helping me produce the volume you hold in your hands. My interest in extinctions was first kindled by Robert H. (Bob) Slaughter of Southern Methodist University, Dallas, Texas. Gerta Keller of Princeton University, New Jersey, has been a long-term collaborator in extinction studies, mostly concerning the end-Cretaceous event, but also across the Late Paleogene transition. David Archibald and Bill Clemens have also been research collaborators and co-conspirators on a number of extinction-related projects. Each of these individuals, along with many others, have contributed to my view of extinction both ancient and modern, but none are responsible for those views or for any part of this text. That responsibility is mine alone. On the production side Colin Ziegler and Trudy Brannan of the Natural History Museum Publishing provided the initial encouragement to develop this book and worked closely with me throughout its production. Thanks go also to Alessandra Serri who worked hard to populate the book with both attractive and informative images.

This book was written largely while in residence at the University of Chicago during the autumn quarter of 2011 on sabbatical leave from the Natural History Museum. Thanks go to Richard Lane for making that sabbatical possible at the Museum end and to both my hosts, W.J.T. Mitchell (Gaylord Donnelley Distinguished Service Professor of English and Art History, University of Chicago) and Julie Lemon (Arts|Science Initiative, Office of the Provost), at the University of Chicago end. The Chicago residency was generously supported by the Franke Institute for the Humanities, the Fishbein Center for the History of Science, the Department of Art History, the University Arts Council, the Nicholson Center for British Studies and the Computation in Science Seminar Series, all of the University of Chicago.

My sincere thanks go to all those mentioned above and to all the many colleagues, researchers, students, reporters and members of the general public (far to numerous to name) with whom I've discussed and debated extinction studies issues over the years. Finally, thanks go to my wife Cecilia, my daughter Jennifer, my friends and my pets (also too numerous to name) who have endured, and I hope will continue to endure, my own obsession with this subject.